Lecture Notes in Computer Science 2233

Edited by G. Goos, J. Hartmanis, and J. van Leeuwen

W0042391

Springer
Berlin
Heidelberg
New York
Barcelona
Hong Kong
London
Milan
Paris
Tokyo

Jon Crowcroft Markus Hofmann (Eds.)

Networked
Group Communication

Third International COST264 Workshop, NGC 2001
London, UK, November 7-9, 2001
Proceedings

 Springer

Series Editors

Gerhard Goos, Karlsruhe University, Germany
Juris Hartmanis, Cornell University, NY, USA
Jan van Leeuwen, Utrecht University, The Netherlands

Volume Editors

Jon Crowcroft
University of Cambridge, Computer Laboratory
William Gates Building, J.J. Thomson Avenue
Cambridge CB3 0FD, United Kingdom
E-mail:jon.crowcroft@cl.cam.ac.uk

Markus Hofmann
Bell Labs/Lucent Technologies
Room 4F-513, 101 Crawfords Corner Road
Holmdel, NJ 07733, USA
E-mail: hofmann@bell-labs.com

Cataloging-in-Publication Data applied for

Die Deutsche Bibliothek - CIP-Einheitsaufnahme

Networked group communication : third international COST 264 workshop /
NGC 2001, London, UK, November 7 - 9, 2001. Jon Crowcroft ; Markus Hofmann
(ed.). - Berlin ; Heidelberg ; New York ; Barcelona ; Hong Kong ; London ;
Milan ; Paris ; Tokyo : Springer, 2001
    (Lecture notes in computer science ; Vol. 2233)
    ISBN 3-540-42824-0

CR Subject Classification (1998):C.2, D.4.4, H.4.3, H.5.3

ISSN 0302-9743
ISBN 3-540-42824-0 Springer-Verlag Berlin Heidelberg New York

Springer-Verlag Berlin Heidelberg New York
a member of BertelsmannSpringer Science+Business Media GmbH

http://www.springer.de

© Springer-Verlag Berlin Heidelberg 2001
Printed in Germany

Typesetting: Camera-ready by author, data conversion by Christian Grosche, Hamburg
Printed on acid-free paper   SPIN 10845787   06/3142   5 4 3 2 1 0

Preface

This year's International Workshop on Networked Group Communications (NGC) was the third event in the NGC series following successful workshops in Pisa, Italy (1999), and Stanford, CA, USA (2000).

For the technical program this year, we received 40 submissions from both academia and industrial institutions all around the world. For each paper, we gathered at least three (and sometimes as many as five) reviews. After a thorough PC meeting that was held in London, 14 papers were selected for presentation and publication in the workshop proceedings. The program continues the themes of previous years, ranging from applications, through security, down to group management, topological considerations, and performance.

The program committee worked hard to make sure that the papers were not only thoroughly reviewed, but also that the final versions were as up to date and as accurate as could be managed in the usual circumstances.

This year, the workshop was held in the London Zoo, hosted by University College London (the tutorials having been held at UCL the day before the start of the paper sessions). The proceedings are published by Springer-Verlag in the LNCS series.

There are signs that the scope of the NGC workshops is broadening to include hot topics such as group communication and content networking. We expect to see more submissions in these areas, as well as other new topics, for NGC 2002, which is to be held at the University of Massachusetts at Amhurst, chaired by Brian Levine and Mostafa Ammar.

We hope that the research community will find these proceedings interesting and helpful and that NGC will continue to be an active forum for research on networked group communication and related areas for years to come.

November 2001   Jon Crowcroft and Markus Hofmann

Workshop Co-chairs

Jon Crowcroft   University of Cambridge
Markus Hofmann   Bell Labs

Program Committee

Kevin Almeroth   UC Santa Barbara
Mostafa Ammar   Georgia Institute of Technology
Supratik Bhattacharyya   Sprint ATL
Saleem Bhatti   University College London
Ernst Biersack   Institut Eurécom
Kenneth Birman   Cornell University
Bob Briscoe   BT
Dah Ming Chiu   Sun Microsystems
Jon Crowcroft   University of Cambridge
Walid Dabbous   INRIA
Christophe Diot   Sprint ATL
Jordi Domingo-Pascual   Universitat Politècnica de Catalunya
Wolfgang Effelsberg   University of Mannheim
Serge Fdida   LIP6
JJ Garcia-Luna   UC Santa Cruz
Mark Handley   ACIRI
Markus Hofmann   Bell Labs
Hugh Holbrook   Stanford/Cisco Systems
David Hutchison   Lancaster University
Roger Kermode   Motorola
Peter Key   Microsoft Research
Isidor Kouvelas   Cisco Systems
Guy Leduc   Université de Liège
Helmut Leopold   Telekom Austria
Brian Levine   University of Massachusetts
Jorg Liebeherr   University of Virginia
Laurent Mathy   Lancaster University
Huw Oliver   Hewlett Packard
Colin Perkins   USC/ISI
Radia Perlman   Sun Microsystems
Luigi Rizzo   Università di Pisa
Clay Shields   Purdue University
Ralf Steinmetz   TU Darmstadt
Burkhard Stiller   ETH Zurich
Don Towsley   University of Massachusetts
Lorenzo Vicisano   Cisco Systems
Brian Whetten   Talarian
Hui Zhang   Carnegie Mellon University

Local Arrangements Chair

Saleem Bhatti   University College London

Local Organization

Jon Crowcroft   University of Cambridge
Saleem Bhatti   University College London
Tristan Henderson   University College London

Supporting/Sponsoring Societies

COST 264
ACM SIGCOMM

Supporting/Sponsoring Companies

Table of Contents

Application-Level

Latency and User Behaviour on a Multiplayer Game Server 1
   Tristan Henderson (University College London)

Application-Level Multicast Using Content-Addressable Networks 14
   Sylvia Ratnasamy (University of California, Berkeley and ACIRI),
   Mark Handley (ACIRI), Richard Karp (University of California,
   Berkeley and ACIRI), Scott Shenker (ACIRI)

SCRIBE: The Design of a Large-Scale Event Notification Infrastructure 30
   Antony Rowstron, Anne-Marie Kermarrec, Miguel Castro (Microsoft
   Research), Peter Druschel (Rice University)

Group Management

SCAMP: Peer-to-Peer Lightweight Membership Service for Large-Scale Group
Communication 44
   Ayalvadi J. Ganesh, Anne-Marie Kermarrec, Laurent Massoulié
   (Microsoft Research)

Extremum Feedback for Very Large Multicast Groups 56
   Jörg Widmer (University of Mannheim),
   Thomas Fuhrmann (Boston Consulting Group)

An Overlay Tree Building Control Protocol 76
   Laurent Mathy (Lancaster University), Roberto Canonico (University
   Federico II, Napoli), David Hutchison (Lancaster University)

Performance

The Multicast Bandwidth Advantage in Serving a Web Site 88
   Yossi Azar (Tel Aviv University), Meir Feder (Bandwiz and Tel Aviv
   University), Eyal Lubetzky, Doron Rajwan, Nadav Shulman (Bandwiz)

STAIR: Practical AIMD Multirate Multicast Congestion Control 100
   John Byers, Gu-In Kwon (Boston University)

Impact of Tree Structure on Retransmission Efficiency for TRACK 113
   Anthony Busson, Jean-Louis Rougier, Daniel Kofman
   (Ecole Nationale Superieure des Telecommunications)

Security

Framework for Authentication and Access Control of Client-Server Group
Communication Systems 128
    Yair Amir, Cristina Nita-Rotaru, Jonathan R. Stanton
    (John Hopkins University)

Scalable IP Multicast Sender Access Control for Bi-directional Trees...... 141
    Ning Wang, George Pavlou (University of Surrey)

EHBT: An Efficient Protocol for Group Key Management 159
    Sandro Rafaeli, Laurent Mathy, David Hutchison
    (Lancaster University)

Topology

Aggregated Multicast with Inter-Group Tree Sharing 172
    *Aiguo Fei, Junhong Cui, Mario Gerla (University of California, Los
Angeles), Michalis Faloutsos (University of California, Riverside)*

Tree Layout for Internal Network Characterizations in Multicast Networks 189
    *Micah Adler, Tian Bu, Ramesh K. Sitaraman, Don Towsley
(University of Massachusetts)*

Author Index 205

Latency and User Behaviour on a Multiplayer Game Server

Tristan Henderson*

Department of Computer Science, University College London
Gower Street, London WC1E 6BT, UK
T.Henderson@cs.ucl.ac.uk

Abstract. Multiplayer online games represent one of the most popular forms of networked group communication on the Internet today. We have been running a server for a first-person shooter game, Half-Life. In this paper we analyse some of the delay characteristics of different players on the server and present some interim results. We find that whilst network delay has some effect on players' behaviour, this is outweighed by application-level or exogenous effects. Players seem to be remarkably tolerant of network conditions, and absolute delay bounds appear to be less important than the relative delay between players.

1 Introduction

Multiplayer online games represent one of the most popular forms of networked group communication on the Internet today, and they contribute to an increasingly large proportion of network traffic [11]. There has been little work to analyse or characterise these applications, or to determine any specific user or network requirements. The real-time nature of many of these games means that response times are important, and in a networked environment this means that round-trip delays must be kept to a minimum. Is network delay, however, the most important factor in a user's gaming experience? In this paper we examine the relationship between application-level delay and player behaviour in multiplayer networked games. The main question that we wished to answer was "How important is delay in a player's decision to select and stay on a particular games server?". To achieve this, we have been running a publicly-accessible server for one of the more popular FPS (First Person Shooter) games, Half-Life [13]. From this server, we have logged and analysed usage behaviour at both the application and network level. The paper is structured as follows. In Section 2 we look at previous work and discuss some expectations we had prior to this study. Section 3 describes our server setup and data collection methodology. Section 4 describes the results that we observed, and Section 5 outlines directions for further work.

---

* The author is funded by a Hewlett-Packard EPSRC CASE award.

J. Crowcroft and M. Hofmann (Eds.): NGC 2001, LNCS 2233, pp. 1–13, 2001.

2   Background

In this section we discuss previous work on multiplayer games, and what this led us to expect before commencing this study.

Previous analysis of popular commercial networked games has thus far concentrated on observing local area network traffic and behaviour [2,4], and network-level rather than user- and -session-level characteristics. There is also little empirical analysis of the delay requirements for real-time multiplayer applications. However, it is generally accepted that low latencies are a requirement. Cheshire [5] proposes 100ms as a suitable bound, whilst human factors research indicates that 200ms might be a more appropriate limit [1]. The IEEE DIS (Distributed Interactive Simulation) standard stipulates a latency bound of between 100ms and 300ms for military simulations [8]. MacKenzie and Ware find that in a VR (Virtual Reality) environment, interaction becomes very difficult above a delay of 225ms [10]. Such previous work implies that there should be an absolute bound to player delay, beyond which players' gameplay becomes so impaired that they would either abort the game or find another server.

MiMaze was a multiplayer game which ran over the multicast backbone (MBone). Some statistics related to a network session of the game are presented in [6]. Using a sample of 25 players, they find that the average client delay is 55ms, and as such the state synchronisation mechanisms are designed with a maximum delay of 100ms in mind. The limited nature of the MBone, however, means that the measurements taken might not be representative of games today. The average delay is far lower than what we observe, and might result from the fact that all clients were located within the same geographic region (France) and well-connected to each other via the MBone.

Like MiMaze, Half-Life uses a client-server architecture, but the distribution model is unicast. Players connect to a common server, which maintains state about the nature of the game. The objective of the game is to shoot and kill as many of the other players as possible. This setup is representative of most of the popular FPS games. Games are typically small, between 16 and 32 players, and there are several thousand servers located all over the Internet. Table 1 shows the average number of servers for some of the more popular games. We obtained these figures by querying master servers for these games every 12 hours for two months. Since there are lots of small groups on servers located all over the Internet, it is reasonable to assume that players will attempt to connect to one of the servers with the lowest delay. This implies that most players would come from geographic locations near to our server, assuming that these have lower delays.

The nature of FPS games means that delay should be an important factor in the gaming experience. Players run around a large "map" (the virtual world) picking up weapons and firing at each other. Low response times should therefore be an advantage, since players can then respond more successfully to other users. If, however, the main benefit of low delay is to gain an advantage over other players, then we might expect that absolute delay is not so important as the

Table 1. Average Number of Servers for Different FPS Games.

Game	Average number of servers
Half-Life	15290.52
Unreal Tournament	2930.16
Quake III Arena	2217.93
Quake II	1207.43
Tribes 2	968.67
QuakeWorld	389.28
Quake	84.94
Sin	49.05
Tribes	42.51
Kingpin	42.20
Heretic II	9.12

variance of delay, if a player needs only to have a relatively low delay compared to the other players in the game.

Usability analysis of non-networked computer games, e.g. [9] indicates that many actions become routine for players as they become expert at the tasks involved, but whether this still holds when the opponents are less predictable (i.e., other humans) is unclear. Unfortunately, we know of no usability studies that specifically analyse networked games, but perhaps regular and highly-skilled players might be able to better tolerate delay, as they become more adept at the specific details and strategy of a particular game, such as Half-Life.

3   Methodology

We recorded users than connected to a games server that we set up at University College London (UCL) in the UK. This server comprised a 900MHz AMD Athlon PC with 256 Mb of RAM, running Linux kernel version 2.4.2, and was connected to our departmental network via 100BaseT Ethernet. To prevent the possibility of users being prejudiced by connecting to an academic site, we registered a non-geographic .com domain name to use instead of a cs.ucl.ac.uk address. The server was advertised to potential players only by using the game's standard mechanisms, whereby a server registers with a "master server". These master servers exist to provide lists of game servers for players; when a player wishes to play a game, they either connect to a known IP address (obtained through out-of-band information or from previous games), or they query the master server to find a suitable game server.

The game was set to rotate maps every 60 minutes, so as to keep the game interesting for existing and potential players. In addition, players were permitted to vote for the next map or to extend the current map at each map rotation interval. The number of players permitted to access the server was arbitrarily set to 24; although the game can potentially support a much higher number of players, most of the more popular maps only effectively scale to 24 players due to a lack of "spawn points" (locations where players can enter the map). There were no specific game-based sessions or goals imposed; players were free to join and leave the server at any time.

Player behaviour was monitored at both the application and the network level. For application-level logging, we took advantage of the server daemon's built-in logging facilities, and augmented this with an additional third-party server management tool to provide more comprehensive logs. Packet-level monitoring used tcpdump, which was set to log UDP packet headers only.

The data that is analysed here derives from running the server between 21 March 2001 18:33 GMT and 15 April 2001 08:28 BST. In this time we observed 31941 sessions (a single user joining and leaving the server).

3.1   Determining Unique Users

Many of the issues that we examine in Section 4 require knowledege of which sessions correspond to which particular users, for example, examining the average delay across all of a particular players' sessions. Such persistent user/session relationships cannot be determined by network-level traces alone, and session-level data is required. However, the nature of most FPS games, where any user can connect to any appropriate server with a minimal amount of authentication, means that determining which sessions belong to which users can be difficult.

Connecting to a Half-Life server is a two-stage process. The client first authenticates with the so-called "WON Auth Server" (the acronym WON stands for World Oppponent Network, the organisation that runs the gaming website http://www.won.net). The authentication server issues the player with a "WONID", a unique identifier generated using the player's license key, which is provided with the CD-ROM media when a player purchases the Half-Life software. There is thus one unique WONID for each purchased copy of the game. Once a WONID has been generated, the player can connect to the Half-Life server of their choice.

Unfortunately, using the WONIDs as a means of identifying unique players proved insufficient. We observed a large number of duplicate WONIDs, indicated by simultaneous use of the same WONID, or players with the same WONID connecting from highly geographically dispersed locations. This duplication of WONIDs is probably due to the sharing of license keys or the use of pirate copies of the game, and so the same WONID can represent more than one user. This situation is exacerbated because the game server program does not reject multiple users with the same WONID from playing simultaneously (this occurred 493 times during the period of this study). In addition, on two occasions the WON Authentication Server seemed to malfunction, issuing all users with a WONID of 0. Although it would have been possible to modify the server to reject simultaneous duplicate WONIDs, this would not resolve the problem of different players connecting at different times with the same WONID, and so we needed to try and determine which sessions belonged to which different individuals.

The identifying information logged by the server for each player is the player's WONID, their IP address and port number, and the nickname that they choose to use in the game. Of the 14776 total WONIDs that we observed, 11612 had unique (WONID, nickname, IP address, port) tuples; we can be reasonably sure that each of these represents one unique user. Of the remaining WONIDs, we had to make some assumptions about users in order to determine their uniqueness. We assume that a unique (WONID,nickname) tuple across all sessions is a single user, and probably has a dynamically-assigned IP address or multiple access ISPs. Users tend to change their names quite often, both between and during sessions, and by looking at all the names used by a particular WONID and looking for common names it was possible to further isolate potential unique users. When multiple users with the same WONID were simultaneously connected we assume that these represent different players. Using these heuristics, we estimate that the 14776 WONIDs actually represent 16969 users.

3.2   Measuring Delay

To measure the delay observed by each player we used the game server's built-in facilities. The server was set up to log the application-level round-trip delay every 30 seconds. Of the total 1314007 measurements, we removed the 16306 with a value of 0, assuming that they are errors. We also saw 10592 measurements greater than 1000ms, with a maximum of 119043ms. We remove these measurements, also assuming that they are errors, since it is unlikely that any user would be able to play a networked game effectively with a 2 minute delay. Moreover, a similar FPS game, Quake III Arena, also assumes that client delays over 1000ms are errors, but chooses not to report them to users, and so we do the same here.

We did not measure network-level delay, e.g. through ICMP pings, since one of our experimental design criteria was that we did not want to alter the server in any way, or send additional traffic to players, in case this altered player behaviour or deterred some potential players from connecting to the server. With over 15,000 other potential servers to choose from, we did not wish to alter conditions in such a way that players might be driven elsewhere. In any case, it is the application-level delay which the users themselves observe and thus one would expect that this would have a greater effect on their behaviour than network-level delay. Informal testing showed us that on a lightly-loaded client, the delays reported by Half-Life are within 5ms of a network-level ping, but, unsurprisingly, this difference rises with client and server load. Unfortunately, without access to the source code of the game, we cannot be sure what causes the erroneous ($\geq$ 1000ms) delays.

Although Half-Life does include features for refusing players admission depending on their delay, and for compensating for variations in player delays [3], these were disabled on our test server, since these might influence any results concerning relative delays.

In addition to measuring the application-level delay, we also performed a whois lookup on players' IP addresses in order to obtain some indication of their area of origin. Further details of the methodology and results of this analysis, and anonymised versions of the server logs, can be found at the author's webpage (http://www.cs.ucl.ac.uk/staff/T.Henderson).

4   Results

Our main results can be summarised as follows:

- There is a wide distribution in players' average delay, with over 40% of players experiencing delays of over 225ms.
- Delay does not appear to play a part in a player's decision to return to the server, or to stay on the server.
- Most players connect during 1800-2400, according to their respective time-zones.
- There is some correlation between a player's ability and their delay and session duration.
- Social bonds do not appear to have an effect on player behaviour.

4.1   Absolute Delay

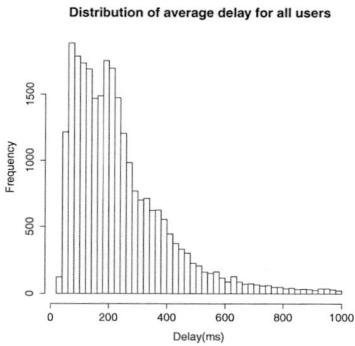

Fig. 1. Distribution of Players' Average Delay.

Figure 1 shows the distribution of the average delay observed over the du-ration of each player's session. The largest proportion of players appear to have delays of between 50 and 300ms, while 95% of players have delays under 533ms (Table 2). The large number of users with high delays of over 225ms is interest-ing since gameplay should theoretically be quite difficult at this level. However, Figure 1 also includes the delays of "tourists"; those players who connect to a server, examine the status of the game and then choose to leave. Figure 2(a) shows the distribution of delay for all the players compared to those who stay less than a minute, and those who stay more than 10 minutes and 1 hour. It can be seen that the delay of those players who stay less than a minute is generally higher than those who stay for longer. A player with a delay of over 400ms is 2.68 times as likely to stay for less than one minute. This implies that delay is a determinant in a player's decision to join a server; players with high delays to a particular server will look elsewhere.

If, however, 100-225ms represents the upper delay bounds for interaction, then we would expect that most of the players with delay above this level would choose other servers. Yet 40.56% of the players who stay for more than one minute, have average delays of over 225ms, and there is no significant difference in the duration of players with delays over 225ms.

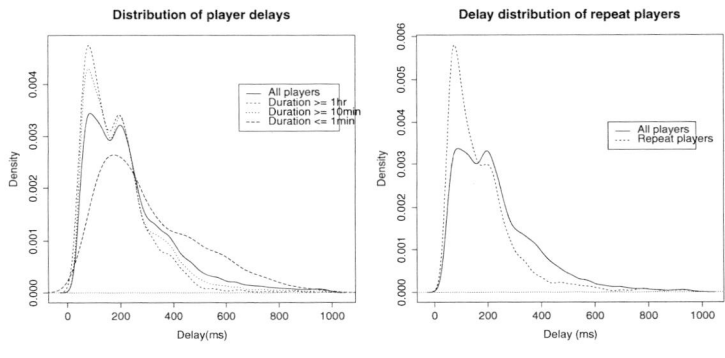

(a) Distribution of Players' Average Delay.

(b) Distribution of Repeat Players' Delay.

Fig. 2. Kernel Density Functions of Delay Distribution.

We define regular players as those who played more than 10 times and whose average session duration exceeded one minute. There were 279 such players. Figure 2(b) indicates that the repeat players' mean delay tends to be lower, but this is statistically insignificant at a 5% confidence level.

Table 2. Overall Delay Results.

Players	Mean delay (ms)	95th percentile
All	232	533
Regular	176	424
Tourists	339	733

4.2   Relative Delay

Absolute delay bounds might not be that important because players become accustomed to high delays, or they have no choice because they happen to have poor network connectivity. A more important delay metric might be the relative delay between players. If one player has a much lower delay than the other players in the game, they might be able to exploit this advantage, by attacking players before they are able to respond.

We measure the relative delay in two ways. First, we look at the nominal results and calculate the standard deviation of players' delay to give an estimate of range. Secondly, we analyse the ordinal data and look at a player's rank in terms of delay compared to the other players.

For most of the time, there is a deviation of around 100ms between players (Figure 3). This seems reasonable, given that most players' delay is in the region of 100-200ms.

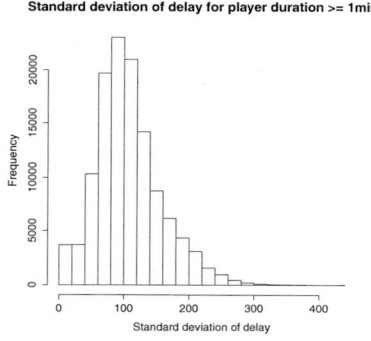

Fig. 3. Distribution of Standard Deviation of Delay.

The "delay rank" of a particular player was calculated by ordering the players at each delay measurement period by delay to produce a rank r, and then scaling the number of players n against the potential maximum number of players of 24, i.e. $r \times 24/n$. Thus, the player with the highest delay would always receive a delay rank of 24, whereas the minimum possible score of 1 would only be possible if the player had the lowest delay and there were 24 players on the server. Figure 4 indicates some correlation between the delay ranks; players who leave tend to have a higher rank.

4.3   Leaving a Game

If delay is a determinant of user behaviour, then one might expect to see a change in delay towards the end of a user's session. A sudden increase in delay might lead a user to leave the server and connect to another one, or give up playing

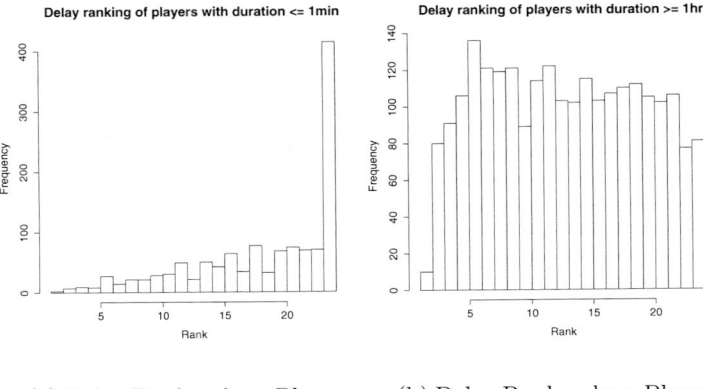

(a) Delay Ranks where Player Duration ≤ 1min.

(b) Delay Ranks where Player Duration ≥ 1hr.

Fig. 4. Relative Delay Ranks.

altogether. We see little evidence for this hypothesis, however. Figure 5(a) shows the "exit delay" (the delay over the last 5% of a player's session) compared to the average delay over the length of the session. This ratio congregates around the value of 1; i.e., the exit delay is usually comparable to the average delay.

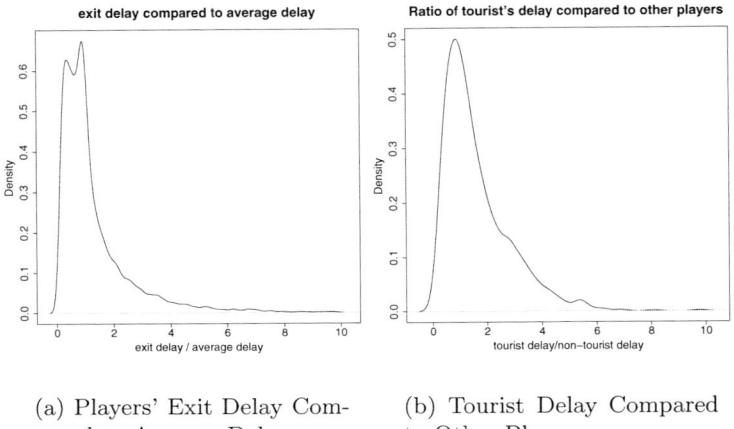

(a) Players' Exit Delay Compared to Average Delay.

(b) Tourist Delay Compared to Other Players.

Fig. 5. Exit and Tourists' Delay.

Relative delay does not seem to be a determinant of tourists leaving the server, either. Figure 5(b) shows the ratio of the delay of those tourists who join and leave the server, compared to the delay of the players already on the server. The mean is 1.618, and there is no correlation between the two.

4.4   When Do Players Play?

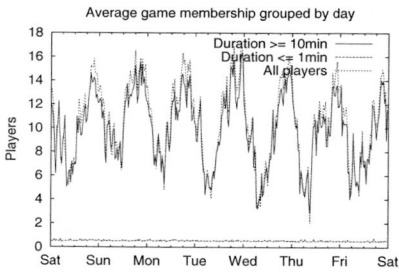
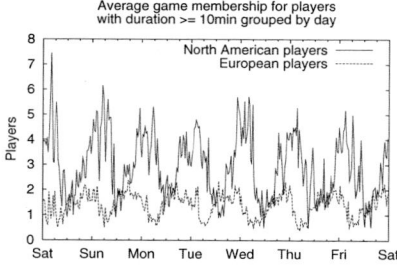

(a) Average Number of Players Each Day.

(b) Average Number of North American and European Players Each Day with Duration ≥ 10 min.

Fig. 6. Number of Players over Time.

Figure 6(a) shows the average number of players for each day of the week, sampled every thirty minutes. There is a strong time-of-day component, which agrees with our previously-observed results [7]. This is perhaps surprising given that players come from areas with different time zones; Figure 6(b) shows the number of players from Europe and North America (determined from the whois database). The offset in the respective peak times is probably due to time zones. Unsurprisingly, the peak usage times are in the late evening, from around 1800 to 2400. If users tend to play in their spare time, then perhaps they have already allocated this time for gameplay, and so are willing to put up with whatever network conditions they happen to experience.

4.5   Player Skill

A player's ability might have an effect on the delay that they can tolerate. A user who is highly skilled at playing the game might be able to cope with higher delays than beginners, since they might be able to predict other player's behaviour and thus compensate for higher-than-average lag. In human factors terms, a high level of skill might lead to players being able to perform actions without conscious awareness — in other words, playing the game becomes automatic.

The session-level logs include details of which players killed each other, and with which particular weapon. Using this information it is possible to estimate of the skill of each player. Whenever a kill takes place, we calculate the players' skill using the following formula:

$$s_k = s_k + s_k/s_d * w; s_d = s_d - s_k/s_d * w$$

where s_k = killer's skill, s_d = killed player's skill, and w is an adjustment for the weapon used (e.g., it is harder to kill with a crowbar than a machine gun).

Using this metric, we see no correlation between skill and delay, nor skill and the duration of a player's game. We see some positive correlation ($R = 0.378$) between session duration and skill, so the more expert players do tend to stay longer. There is a slight negative correlation ($R = -0.231$) between skill and delay, so a lower delay may lead to improved performance.

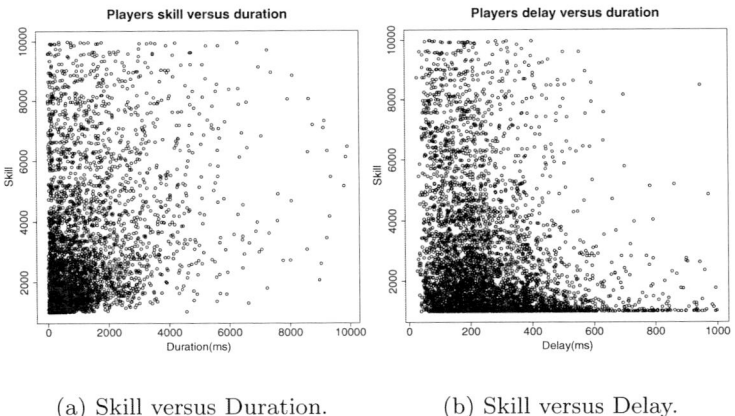

(a) Skill versus Duration.   (b) Skill versus Delay.

Fig. 7. Effect of Skill.

4.6   Social Bonds

Figure 2(b) indicates the presence of certain players who had excessively high delays, yet kept returning to the server. Here we analyse some of these players in more detail to see if it is possible to determine why they keep coming back. Table 3 shows some of these players' statistics.

One possibility is that these players are returning to the server because of friends who are also playing. However, of the 283 other players who were on the

Table 3. Detailed Statistics for Regular Players with High Delays.

TLD (from whois)	Number of visits	Average delay (ms)	Maximum delay (ms)	Average duration (sec)	Maximum duration (sec)
TW	10	423	533	390	1460
TH	38	794	953	2319	10418
KR	14	340	389	2477	10043
TW	15	554	771	665	1659
KR	13	339	362	970	2553

server at the same time as these five players, only 33 players appear twice and one appears three times. It is therefore unlikely that repeat visits or social bonds were the reason for these players returning.

5   Conclusions, Caveats, and Future Work

This study has looked at the effects of delay on user dynamics on a multiplayer games server. We find that application-level delay does not appear to have a significant effect on a user's behaviour once they have chosen to connect to a games server. Although the majority of users have delays within the bounds predicted by previous VR and DIS studies, changes in this delay do not seem to lead to players aborting the game.

Although brief and still at an interim stage, this study has raised a number of interesting research questions. If users are concerned with relative delay, is it possible to design efficient algorithms for determining the server with the smallest standard deviation in delay given a group of prospective users? There is currently a lot of interest in optimal placement of web and mirror servers e.g. [12] and perhaps this could be extended to locating game servers.

That games players might be unaffected by sudden changes in delay has important implications for designing potential congestion control schemes for games. In particular, pricing schemes that depend on users adapting to network conditions because of price changes might be less practical. If users tend to remain in a game for an exogenously determined duration, then session-based pricing or reservations might make more sense from a user's perspective. Pricing schemes for games could be designed to adapt the network to the user (who has already committed to playing a game), rather than the other way around.

Usability studies of multiplayer networked FPS games, e.g. a GOMS (Goals, Operators, Methods and Selection) analysis such as that performed in [9], might help to explain some of the results we have seen, since simple correlations of kills and deaths appear to be insufficient. More elaborate skill metrics, e.g. taking into account the amount of time between deaths, might also prove fruitful.

As our results are only from the study of a single Half-Life server, which may not be representative of servers across the Internet as a whole, we intend to investigate lightweight methods for instrumenting larger numbers of servers for future data collection. In spite of these limitations, this study has provided us with some direction for future experimental work. This study has only been correlational — we have examined and attempted to interpret results from an unmodified server. In future work we intend to run multiple servers, modifying variables such as network delay and jitter, and simulating different congestion control and QoS policies, to further investigate their effects on user behaviour. We expect to find our study corroborated by this further study.

Acknowledgement

Thanks to Saleem Bhatti, Jon Crowcroft, and the reviewers for their comments.

References

1. R. W. Bailey. *Human Performance Engineering — Using Human Factors/Ergonomics to Achieve Computer System Usability*. Prentice Hall, Englewood Cliffs, NJ, second edition, 1989.
2. R. A. Bangun, E. Dutkiewicz, and G. J. Anido. An analysis of multi-player network games traffic. In *Proceedings of the 1999 International Workshop on Multimedia Signal Processing*, pages 3–8, Copenhagen, Denmark, Sept. 1999.
3. Y. W. Bernier. Latency compensating methods in client/server in-game protocol design and optimization. In *Proceedings of the 15th Games Developers Conference*, San Jose, CA, Mar. 2001.
4. M. S. Borella. Source models of network game traffic. *Computer Communications*, 23(4):403–410, Feb. 15, 2000.
5. S. Cheshire. Latency and the quest for interactivity, Nov. 1996. White paper commissioned by Volpe Welty Asset Management, L.L.C., for the Synchronous Person-to-Person Interactive Computing Environments Meeting.
6. L. Gautier and C. Diot. Design and evaluation of MiMaze, a multi-player game on the Internet. In *Proceedings of the 1998 IEEE International Conference on Multimedia Computing and Systems*, pages 233–236, Austin, TX, June 1998.
7. T. Henderson and S. Bhatti. Modelling user behaviour in networked games. In *Proceedings of the 9th ACM Multimedia Conference*, Ottawa, Canada, Oct. 2001.
8. Institute of Electrical and Electronic Engineers. *1278.2-1995, IEEE Standard for Distributed Interactive Simulation — Communication Services and Profiles*. IEEE, New York, NY, Apr. 1996.
9. B. E. John and A. H. Vera. A GOMS analysis of a graphic, machine-paced, highly interactive task. In *Proceedings of the CHI'92 Conference on Human factors in computing systems*, pages 251–258, Monterey, CA, May 1992.
10. I. S. MacKenzie and C. Ware. Lag as a determinant of human performance in interactive systems. In *Proceedings of the CHI '93 Conference on Human factors in computing systems*, pages 488–493, Amsterdam, The Netherlands, Apr. 1993.
11. S. McCreary and K. Claffy. Trends in wide area IP traffic patterns: A view from Ames Internet Exchange. In *Proceedings of the ITC Specialist Seminar on IP Traffic Modeling, Measurement and Management*, Monterey, CA, Sept. 2000.
12. L. Qiu, V. N. Padmanabhan, and G. M. Voelker. On the placement of web server replicas. In *Proceedings of the 20th IEEE Conference on Computer Communications (INFOCOM)*, pages 1587–1596, Anchorage, AK, Apr. 2001.
13. Valve Software. Half-Life. http://www.sierrastudios.com/games/half-life/.

Application-Level Multicast Using Content-Addressable Networks

Sylvia Ratnasamy[1,2], Mark Handley[2], Richard Karp[1,2], and Scott Shenker[2]

[1] University of California, Berkeley, CA, USA
[2] AT&T Center for Internet Research at ICSI

Abstract. Most currently proposed solutions to application-level multicast organise the group members into an application-level mesh over which a Distance-Vector routing protocol, or a similar algorithm, is used to construct source-rooted distribution trees. The use of a global routing protocol limits the scalability of these systems. Other proposed solutions that scale to larger numbers of receivers do so by restricting the multicast service model to be single-sourced. In this paper, we propose an application-level multicast scheme capable of scaling to large group sizes without restricting the service model to a single source. Our scheme builds on recent work on Content-Addressable Networks (CANs). Extending the CAN framework to support multicast comes at trivial additional cost and, because of the structured nature of CAN topologies, obviates the need for a multicast routing algorithm. Given the deployment of a distributed infrastructure such as a CAN, we believe our CAN-based multicast scheme offers the dual advantages of simplicity and scalability.

1 Introduction

Several recent research projects[8,10,7] propose designs for application-level networks wherein nodes are structured in some well-defined manner. A Content-Addressable Networks (CANs)[6] is one such system. Briefly,[1] a Content-Addressable Network is an application-level network whose constituent nodes can be thought of as forming a virtual d-dimensional Cartesian coordinate space. Every node in a CAN "owns" a portion of the total space. For example, Figure 1 shows a 2-dimensional CAN occupied by 5 nodes. A CAN, as described in [6], is scalable, fault-tolerant and completely distributed. Such CANs are useful for a range of distributed applications and services. For example, in [6] we focus on the use of a CAN to provide hash table-like functionality on Internet-like scales – a function useful for indexing in peer-to-peer applications, large-scale storage management systems, the construction of wide-area name resolution services and so forth.

This paper looks into the question of how the deployment of such CAN-like distributed infrastructures might be utilised to support multicast services and applications. We outline the design of an application-level multicast scheme built

---

[1] Section 2 describes the CAN design in some detail.

J. Crowcroft and M. Hofmann (Eds.): NGC 2001, LNCS 2233, pp. 14–29, 2001.

using a CAN. Our design shows that extending the CAN framework to support multicast comes at trivial additional cost in terms of complexity and added protocol mechanism. A key feature of our scheme is that because we exploit the well-defined structured nature of CAN topologies (i.e. the virtual coordinate space) we can eliminate the need for a multicast routing algorithm to construct distribution trees. This allows our CAN-based multicast scheme to scale to large group sizes. While our design is in the context of CANs in particular, we believe our technique of exploiting the structure of these systems should be applicable to the Chord [8], Pastry [7] and Tapestry [10] designs.

In previous work, several research proposals have argued for *application-level* multicast[1,3,4] as a more tractable alternative to a network-level multicast service and have described designs for such a service and its applications. The majority of these proposed solutions (for example [1,4]) typically involve having the members of a multicast group self-organise into an essentially random application-level mesh topology over which a traditional multicast routing algorithm such as DVMRP [2] is used to construct distribution trees rooted at each possible traffic source. Such routing algorithms require every node to maintain state for every other node in the topology. Hence, although these proposed solutions are well suited to their targeted applications,[2] their use of a global routing algorithm limits their ability to scale to large (many thousands of nodes) group sizes and to operate under conditions of dynamic group membership.

Bayeux[11] is an application-level multicast scheme that scales to large group sizes but restricts the service model to a single source. In contrast to the above schemes, CAN-based multicast can scale to large group sizes without restricting the service model to a single source.

In summary, we believe our CAN-based multicast scheme offers two key advantages:

- CAN-based multicast can scale to very large (*i.e.* many thousands of nodes and higher) group sizes without restricting the service model to a single-source. To the best of our knowledge, no currently proposed application-level multicast scheme can operate in this regime.
- Assuming the deployment of a CAN-like infrastructure, CAN-based multicast is trivially simple to achieve. This is not to suggest that CAN-based multicast by itself is either simpler or more complex than other proposed solutions to application-level multicast. Rather, our point is that CANs can serve as a building block in a range of Internet applications and services and that one such, easily achievable, service is application-level multicast.

The remainder of this paper is organised as follows: Section 2 reviews the design and operation of a CAN. We describe the design of a CAN-based multicast service in Section 3 and evaluate this design through simulation in Section 4. Finally, we discuss related work in Section 5 and conclude.

---

[2] The authors in [4], state that End System Multicast is more appropriate for small, sparse groups as in audio-video conferencing and virtual classrooms, while the authors in [1] apply their algorithm, Gossamer, to the self-organisation of infrastructure proxies.

2   Content-Addressable Networks

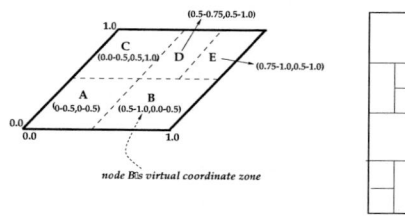

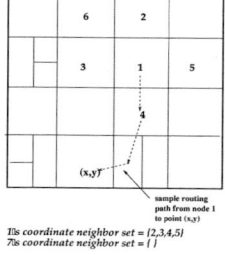

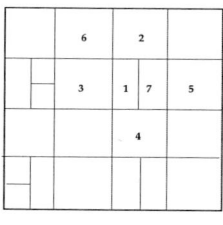

Fig. 1.   *Example 2-d Coordinate Overlay with 5 Nodes.*

Fig. 2.   *Example 2-d Space before Node 7 Joins.*

Fig. 3.   *Example 2-d Space after Node 7 Joins.*

In this Section, we present our design of a Content-Addressable Network. This paper gives only a brief overview of our CAN design; [6] presents the details and evaluation.

2.1   Design Overview

Our design centers around a virtual d-dimensional Cartesian coordinate space on a d-torus. [3] This coordinate space is completely logical and bears no relation to any physical coordinate system. At any point in time, the *entire* coordinate space is dynamically partitioned among all the nodes in the system such that every node "owns" its individual, distinct zone within the overall space. For example, Figure 1 shows a 2-dimensional $[0, 1] \times [0, 1]$ coordinate space partitioned between 5 CAN nodes. This coordinate space provides us with a level of indirection, since one can now talk about storing content at a "point" in the space or routing between "points" in the space where a "point" refers to the node in the CAN that owns the zone enclosing that point.

For example, this virtual coordinate space is used to store (key,value) pairs as follows: to store a pair (K_1, V_1), key K_1 is deterministically mapped onto a point, say (x, y) in the coordinate space using a uniform hash function. The corresponding key-value pair is then stored at the node that owns the zone within which the point (x, y) lies. To retrieve an entry corresponding to key K_1, any node can apply the same deterministic hash function to map K_1 onto point (x, y) and then retrieve the corresponding value from the point (x, y). If the point (x, y) is not owned by the requesting node or its immediate neighbours, the request must be routed through the CAN infrastructure until it reaches the

---

[3] For simplicity, the illustrations in this paper do not show a torus.

node in whose zone (x, y) lies. Efficient routing is therefore a critical aspect of our CAN.

Nodes in the CAN self-organise into an overlay network that represents this virtual coordinate space. A node learns and maintains as its set of neighbours the IP addresses of those nodes that hold coordinate zones adjoining its own zone. This set of immediate neighbours serves as a coordinate routing table that enables routing between arbitrary points in the coordinate space.

We first describe the three most basic pieces of our design: CAN routing, construction of the CAN coordinate overlay, and maintenance of the CAN overlay and then briefly discuss the simulated performance of our design.

2.2   Routing in a CAN

Intuitively, routing in a Content Addressable Network works by following the straight line path through the Cartesian space from source to destination coordinates.

A CAN node maintains a coordinate routing table that holds the IP address and virtual coordinate zone of each of its neighbours in the coordinate space. In a d-dimensional coordinate space, two nodes are neighbours if their coordinate spans overlap along $d-1$ dimensions and abut along one dimension. For example, in Figure 2, node 5 is a neighbour of node 1 because its coordinate zone overlaps with 1's along the Y axis and abuts along the X-axis. On the other hand, node 6 is not a neighbour of 1 because their coordinate zones abut along both the X and Y axes. This purely local neighbour state is sufficient to route between two arbitrary points in the space: A CAN message includes the destination coordinates. Using its neighbour coordinate set, a node routes a message towards its destination by simple greedy forwarding to the neighbour with coordinates closest to the destination coordinates. Figure 2 shows a sample routing path.

For a d dimensional space partitioned into n equal zones, the average routing path length is thus $(d/4)(n^{1/d})$ and individual nodes maintain $2d$ neighbours. These scaling results mean that for a d dimensional space, we can grow the number of nodes (and hence zones) without increasing per node state while the path length grows as $O(n^{1/d})$.

Note that many different paths exist between two points in the space and so, even if one or more of a node's neighbours were to crash, a node would automatically route along the next best available path. If however, a node loses all its neighbours in a certain direction, and the repair mechanisms described in Section 2.4 have not yet rebuilt the void in the coordinate space, then greedy forwarding may temporarily fail. In this case, a node may use an expanding ring search to locate a node that is closer to the destination than itself. The message is then forwarded to this closer node, from which greedy forwarding is resumed.

2.3   CAN Construction

As described above, the entire CAN space is divided amongst the nodes currently in the system. To allow the CAN to grow incrementally, a new node that joins the system must be allocated its own portion of the coordinate space. This is done by an existing node splitting its allocated zone in half, retaining half and handing the other half to the new node.

The process takes three steps:

1. First the new node must find a node already in the CAN.
2. Next, using the CAN routing mechanisms, it must find a node whose zone will be split.
3. Finally, the neighbours of the split zone must be notified so that routing can include the new node.

Bootstrap. A new CAN node first discovers the IP address of any node currently in the system. The functioning of a CAN does not depend on the details of how this is done, but we use the same bootstrap mechanism as Yallcast and YOID [3]. As in [3] we assume that a CAN has an associated DNS domain name, and that this resolves to the IP address of one or more CAN bootstrap nodes. A bootstrap node maintains a partial list of CAN nodes it believes are currently in the system. Simple techniques to keep this list reasonably current are described in [3]. To join a CAN, a new node looks up the CAN domain name in DNS to retrieve a bootstrap node's IP address. The bootstrap node then supplies the IP addresses of several randomly chosen nodes currently in the system.

Finding a Zone. The new node then randomly chooses a point (x, y) in the space and sends a JOIN request destined for point (x, y). This message is sent into the CAN via any existing CAN node. Each CAN node then uses the CAN routing mechanism to forward the message, until it reaches the node in whose zone (x, y) lies.

This current occupant node then splits its zone in half and assigns one half to the new node. The split is done by assuming a certain ordering of the dimensions in deciding along which dimension a zone is to be split, so that zones can be re-merged when nodes leave. For a 2-d space a zone would first be split along the X dimension, then the Y and so on. The (key, value) pairs from the half zone to be handed over are also transfered to the new node.

Joining the Routing. Having obtained its zone, the new node learns the IP addresses of its coordinate neighbour set from the previous occupant. This set is a subset of the the previous occupant's neighbours, plus that occupant itself. Similarly, the previous occupant updates its neighbour set to eliminate those nodes that are no longer neighbours. Finally, both the new and old nodes' neighbours must be informed of this reallocation of space. Every node in the system sends an immediate update message, followed by periodic refreshes, with its currently assigned zone to all its neighbours. These soft-state style updates ensure that all of their neighbours will quickly learn about the change and will update their own neighbour sets accordingly. Figures 2 and 3 show an example of a new node (node 7) joining a 2-dimensional CAN.

As can be inferred, the addition of a new node affects only a small number of existing nodes in a very small locality of the coordinate space. The number of neighbours a node maintains depends only on the dimensionality of the coordinate space and is independent of the total number of nodes in the system. Thus, node insertion affects only *O(number of dimensions)* existing nodes which is important for CANs with huge numbers of nodes.

2.4   Node Departure, Recovery, and CAN Maintenance

When nodes leave a CAN, we need to ensure that the zones they occupied are taken over by the remaining nodes. The normal procedure for doing this is for a node to explicitly hand over its zone and the associated (key,value) database to one of its neighbours. If the zone of one of the neighbours can be merged with the departing node's zone to produce a valid single zone, then this is done. If not, then the zone is handed to the neighbour whose current zone is smallest, and that node will then temporarily handle both zones.

The CAN also needs to be robust to node or network failures, where one or more nodes simply become unreachable. This is handled through a recovery algorithm, described in [6], that ensures one of the failed node's neighbours takes over the zone.

2.5   Design Improvements and Performance

Our basic CAN algorithm as described in the previous section provides a balance between low per-node state ($O(d)$ for a d dimensional space) and short path lengths with $O(dn^{1/d})$ hops for d dimensions and n nodes. This bound applies to the number of hops in the CAN path. These are *application level* hops, not IP-level hops, and the latency of each hop might be substantial; recall that nodes that are adjacent in the CAN might be many miles (and many IP hops) away from each other. In [6], we describe a number of design techniques whose primary goal is to reduce the latency of CAN routing. Of particular relevance to the work in this paper, is a distributed "binning" scheme whereby co-located nodes on the Internet can be placed close by in the CAN coordinate space. In this scheme, every node independently measures its distance (*i.e.* latency) from a set of well known landmark machines and joins a particular portion of the coordinate space based on these measurements. Our simulation results in [6] indicate that these added mechanisms are very effective in reducing overall path latency. For example, we show that for a system with over 130,000 nodes, for a range of link delay distributions, we can route with a latency that is well within a factor of three of the underlying IP network latency. The number of neighbours that a node must maintain to achieve this is approximately 28 (details of this test are in Section 4 in [6]).

3   CAN-Based Multicast

In this section, we describe a solution whereby CANs can be used to offer an application-level multicast service.

If all the nodes in a CAN are members of a given multicast group, then multicasting a message only requires flooding the message over the entire CAN. As we shall describe in Section 3.2, we can exploit the existence of a well defined coordinate space to provide simple, efficient flooding algorithms from arbitrary sources without having to compute distribution trees for every potential source.

If only a subset of the CAN nodes are members of a particular group, then multicasting involves two pieces:

- the members of the group first form a group-specific "mini" CAN and then,
- multicasting is achieved by flooding over this mini CAN

In what follows, we describe the two key components of our scheme: group formation and multicast by flooding over the CAN.

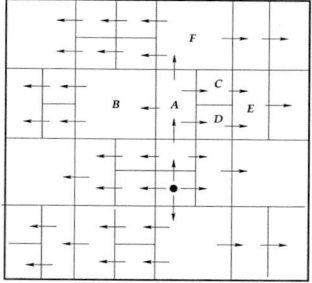

Fig. 4.   *Directed Flooding over the CAN.*

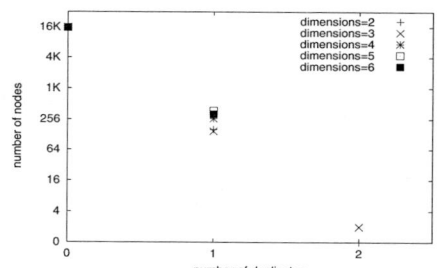

Fig. 5.   *Duplicate Messages Using CAN-Based Multicast.*

3.1   Multicast Group Formation

To assist in our explanation, we assume the existence of a CAN C within which a subset of the nodes wish to form a multicast group G. We achieve this by forming an additional mini CAN, call it C_g, made up of only the members of G. The underlying CAN C itself is used as the bootstrap for the formation of C_g as follows: using a well-known hash function, the group address G is deterministically mapped onto a point, say (x, y), and the node on C that owns the point (x, y) serves as the bootstrap node in the construction of C_g. Joining group G thus reduces to joining the CAN C_g. This is done by repeating the usual CAN construction process with (x, y) as the bootstrap node. Because of

the light-weight nature of the CAN bootstrap mechanisms, we do not expect
the CAN bootstrap node to be overloaded by join requests. If this becomes a
possibility however, one could use multiple bootstrap nodes to share the load
by using multiple hash functions to deterministically map the group name G
onto multiple points in the CAN C; the nodes corresponding to each of these
points would then serve as a bootstrap node for the group G. As with the
CAN bootstrap process, the failure of the bootstrap node(s) does not affect the
operation of the multicast group itself; it only prevents new nodes from joining
the group during the period of failure.

Thus, every group has a corresponding CAN made up of all the group members. Note that with this group formation process a node only maintains state
for those groups for which it is itself a member or for which it serves as the
bootstrap node. For a d-dimensional CAN, a member node maintains state for
$2d$ additional nodes (its neighbours in the CAN), independent of the number of
traffic sources in the multicast group.

3.2   Multicast Forwarding

Because all the members of group G (and no other node) belong to the associated CAN C_g, multicasting to G is achieved by flooding on the CAN C_g.
Different flooding algorithms are conceivable; for example, one might consider a
naive flooding algorithm wherein a node caches the sequence numbers of messages it has recently received. On receiving a new message, a node forwards the
message to all its neighbours (except of course, the neighbour from which it received the message) only if that message is not already in its cache. With this
type of floodcachesuppress algorithm a source can reach every group member
with requiring a routing algorithm to discover the network topology. Such an
algorithm does not make any special use of the CAN structure and could in fact
be run over any application-level topology including a random mesh topology
as generated in [4,1]. The problem with this type of naive flooding algorithm is
that it can result in a large amount of duplication of messages; in the worst case,
a node could receive a single message from each of its neighbours.

 A more efficient flooding solution would be to exploit the coordinate space
structure of the CAN as follows:

Assume that our CAN is a d-dimensional CAN with dimensions $1 \ldots d$. Individual nodes thus have at least $2d$ neighbours; 2 per dimension with one to
move forward and another to move in reverse along each dimension. $i.e.$ for every
dimension i a node has at least one neighbour whose zone abuts its own own in
the forward direction along i and another neighbour whose zone abuts its own
in the reverse direction along i. For example, consider node A in Figure 4: node
B abuts A in the reverse direction along dimension 1 while nodes C and D abut
A in the forward direction along dimension 1.

Messages are then forwarded as follows:

1. The source node ($i.e.$ node that generates a new message) forwards a message
   to all its neighbours

2. A node that receives a message from a neighbour with which it abuts along dimension i forwards the message to those neighbours with which it abuts along dimension $1 \ldots (i-1)$ and the neighbours with which it abuts along dimension i on the opposite side to that from which it received the message. Figure 4 depicts this directed flooding algorithm for a 2-dimensional CAN.

3. a node does not forward a message along a particular dimension if that message has already traversed at least half-way across the space from the source coordinates along that dimension. This rule prevents the flooding from looping round the back of the space.

4. a node caches the sequence numbers of messages it has received and does not forward a message that it has already previously received

For a perfectly partitioned (*i.e.* where nodes have equal sized zones) coordinate space, the above algorithm ensures that every node receives a message exactly once. For imperfectly partitioned spaces however, a node might receive the same message from more than one neighbour. For example, in Figure 4, node E would receive a message from both neighbours C and D.

Certain duplicates can be easily avoided because, under normal CAN operation, every node knows the zone coordinates for each of its neighbours. For example, consider once more Figure 4; nodes C and D both know each others' and node E's zone coordinates and could hence use a deterministic rule such that only one of them forwards messages to E. Such a rule, however, only eliminates duplicates that arise by flooding along the first dimension. The rule works along the first dimension because, *all* nodes forward along the first dimension. Hence even if a node, by applying some deterministic rule, does not forward a message to its neighbour along the first dimension, we know that some other node that does satisfy the deterministic rule will do so. But this need not be the case when forwarding along higher dimensions. Consider a 3-dimensional CAN; if a node by the application of a deterministic rule decides not to forward to a neighbour along the second dimension, there is no guarantee that any node will eventually forward it up along the second dimension because the node that does satisfy the deterministic rule might receive the packet along the first dimension and hence will not forward the message along the second dimension. [4] For example, in Figure 4 let us assume that node A decides (by the use of some deterministic rule) not to forward to node F. Because node C receives the message (from A) along the first dimension, it will not forward the message along the second dimension either and hence node F and the other nodes with Y-axis coordinates in the same range as F, will never receive the message. While the above strategy does not eliminate all duplicates, it does eliminate a large fraction of it because most of the flooding occurs along the first dimension. Hence, we augment the above flooding algorithm with the following deterministic rule used to eliminate duplicates that arise from forwarding along the first dimension:

- let us assume that a node, P, received a message along dimension 1 and that node Q abuts P along dimension 1 in the opposite direction from which P

---

[4] By the second rule in the flooding algorithm.

received the message. Consider the corner C_q of Q's zone that abuts P along dimension 1 and has the lowest coordinates along dimensions $2 \ldots d$. Then, P only forwards the message on to Q, if P is in contact with the corner C_q.

So, for example, in Figure 4, with respect to nodes C and D, the corner under consideration for node E would be the lower, leftmost corner of E's zone. Hence only D (and not C) would forward messages E in the forward direction along the first dimension.

For the above flooding algorithm, we measured through simulation the percentage of nodes that experienced different degrees of message duplication caused by imperfectly partitioned spaces. Figure 5 plots the number of nodes that received a particular number of duplicate messages for a system with 16,384 nodes using CANs with dimensions ranging from 2 to 6. In all cases, over 97% of the nodes receive no duplicate messages and amongst those nodes that do, virtually all of them receive only a single duplicate message. This is a considerable improvement over the naive flooding algorithm wherein *every* node might receive a number of duplicates up to the degree (number of neighbours) of the node.

It is worth noting that the naive flooding algorithm is very robust to message loss because a node can receive a message via any of its neighbours. However, the efficient flooding algorithm is less robust because the loss of a single message results in the breakdown of message delivery to several subsequent nodes thus requiring additional loss recovery techniques. This problem is however, no different than in the case of traditional IP multicast or other application-level schemes where the loss of a packet along a single link results in the packet being lost by all downstream nodes in the distribution tree. With both flooding algorithms, the duplication of messages arises because we do not (unlike most other solutions to multicast delivery) construct a single spanning tree rooted at the source of traffic. However, we believe that the simplicity and scalability gained by not having to run routing algorithms to construct and maintain such delivery trees is well worth the slight inefficiencies that may arise from the duplication of messages.

Using the above flooding algorithm, any group member can multicast a message to the entire group. Nodes that are not group members can also multicast to the entire group by first discovering a random group member and relaying the transmission through this random group member. [5] This random member node can be discovered by contacting the bootstrap node associated with the group name.

4   Performance Evaluation

In this section, we evaluate, through simulation, the performance of our CAN-based multicast scheme. We adopt the performance metrics and evaluation strategy used in [4]. As with previous evaluation studies of application-level multicast

---

[5] Note that relaying in our case is different from relayed transmissions as done in source specific multicast [11] because only transmissions from non-member nodes are relayed and even these can be relayed through any member node.

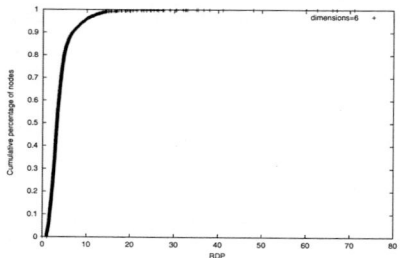

Fig. 6. *Cumulative Distribution of RDP.*

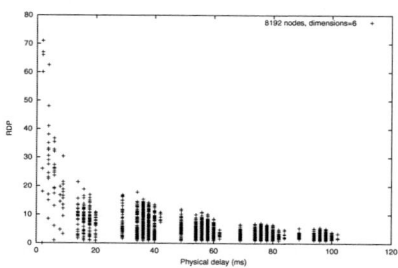

Fig. 7. *RDP versus Physical Delay for Every Group Member.*

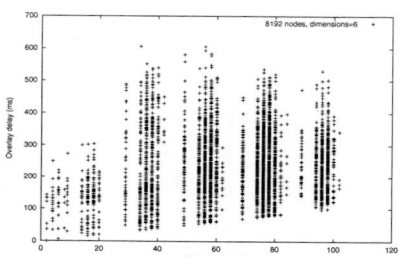

Fig. 8. *Delay on the Overhead versus Physical Network Delay.*

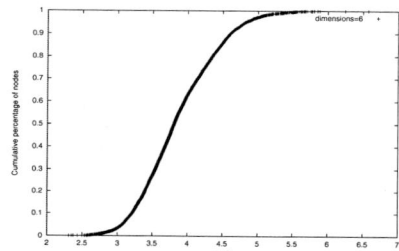

Fig. 9. *Cumulative Distribution of RDP Averaged over 100 Traffic Sources.*

schemes[4,1,11] we compare the performance of CAN-based multicast to native IP multicast and naive unicast-based multicast where the source simply unicasts a message to every receiver in succession. Our evaluation metrics are:

- **Relative Delay Penalty (RDP)** : the ratio of the delay between two nodes (in this case, the source node and a receiver) using CAN-based multicast to the unicast delay between them on the underlying physical network
- **Link Stress** : the number of identical copies of a packet carried by a physical link

Our simulations were performed on Transit-Stub (TS) topologies using the GT-ITM topology generator [9]. TS topologies model networks using a 2-level hierarchy of routing domains with transit domains that interconnect lower level stub domains.

4.1 Relative Delay Penalty

We first present results from a multicast transmission using a single source as this represents the performance typically seen across the different receiver nodes for a transmission from a single source. These simulations were performed using a CAN with 6 dimensions and a group size of 8192 nodes. The source node was selected at random. We used Transit-Stub topologies with link latencies of 20ms for intra-transit domain links, 5ms for stub-transit links and 2ms for intra-stub domain links.

Both IP multicast and Unicast-based multicast achieve an RDP value of one for all group members because messages are transmitted along the direct physical (IP-level) path between the source and receivers. Routing on an overlay network however, fundamentally results in higher delays. Figure 6 plots the cumulative distribution of RDP over the group members. While the majority of receivers see an RDP of less than about 5 or 6, a few group members have a high RDP. This can be explained [6] from the scatter-plot in Figure 7. The figure plots the relation between the RDP observed by a receiver and its distance from the source on the underlying IP-level, physical network. Each point in Figure 7 indicates the existence of a receiver with the corresponding RDP and IP-level delay. As can be seen, all the nodes with high values of RDP have a low physical delay to the source, i.e. the very low delay from these receivers to the source inflates their RDP. However, the absolute value of their delay from the source on the CAN overlay is not really very high. This can be seen from Figure 8, which plots, for every receiver, its delay from the source using CAN multicast versus its physical network delay. The plot shows that while the maximum physical delay can be about 100ms, the maximum delay using CAN-multicast is about 600ms and the receivers on the left hand side of the graph, which had the high RDP, experience delays of not more than 300ms.

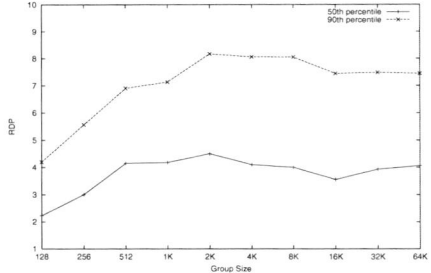

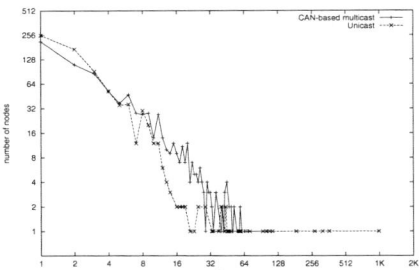

Fig. 10. *RDP versus Increasing Group Size.*

Fig. 11. *Number of Physical Links with a Given Stress.*

---

[6] The authors in [4] make the same observation and explanation.

The above results were all for a single multicast transmission using a single source; Figure 9 plots the cumulative distribution of the RDP with the delays averaged over multicast transmissions from a 100 sources selected at random. Because a node is unlikely to be very close (in terms of physical delay) to all 100 sources, averaging the results over transmissions from many sources helps to reduce the appearance of inflated RDPs that occurs when a receiver is very close to the source. From Figure 9 we see that, on an average, no node sees an RDP of more than about 6.0.

Finally, Figure 10 plots the 50 and 90 percentile RDP values for group sizes ranging from 128 to 65,000 for a single source. We scale the group size as follows: we take a 1,000 node Transit-Stub topology as before and to this topology, we add end-host (source and receiver) nodes to the stub (leaf) nodes in the topology. The delay of the link from the end-host node to the stub node is set to 1ms. Thus in scaling the group size from a 128 to 65K nodes, we're scaling the density of the graph without scaling the backbone (transit) domain. So, for example, a group size of 128 nodes implies that approximately one in ten stub nodes has an associated group member while a group size of 65K implies that every stub node has approximately 65 attached end-host nodes. This method of scaling the graph causes the flat trend in the growth of RDP with group size because for a given source the relative number of close-by and distant nodes stays pretty much constant. Further, at high density, every CAN node has increasingly many close-by nodes and hence the CAN binning technique used to cluster co-located nodes yields higher gains. Different methods for scaling topologies could yield different scaling trends.

While the significant differences between End-System Multicast and CAN-based multicast makes it hard to draw any direct comparison between the two systems; Figure 10 indicates that the performance of CAN-based multicast even for small group sizes is competitive with End-System multicast.

4.2   Link Stress

Ideally, one would like the stress on the different physical links to be somewhat evenly distributed. Using native IP multicast, every link in the network has a stress of exactly one. In the case of unicasting from the source directly to all the receivers, links close to the source node have very high stress (equal to the group size at the first hop link from the source). Figure 11 plots the number of nodes that experienced a particular stress value for a group size of 1024 for a 6-dimensional CAN. Unlike naive unicast where a small number of links see extremely high stress, CAN-based multicast distributes the stress much more evenly over all the links.

Figure 12 plots the worst-case stress for group sizes ranging from 128 to 65,000 nodes. The high stress in the case of large group sizes is because, as described earlier, we scale the group size without scaling the size of the backbone topology. For the above simulation, we used a transit-stub topology with a 1,000 nodes. Hence for a group size of 65,000 nodes, all 65,000 nodes are interconnected by a backbone topology of less than 1,000 nodes thus putting high stress on some backbone links. We repeated the above simulation for a transit-stub topology with 10,000 nodes, thus decreasing the density of the graph by a factor of 10. Figure 13 plots the worst-case stress for group sizes up to 2,048 nodes for all three cases (*i.e.* CAN-based multicast using Transit-Stub topologies with 1,000 and 10,000 nodes and naive unicast-based multicast). As can be seen, at lower density the worst-case stress drops sharply. For example, at 2,048 nodes the worst case stress drops from 169 (for TS1000) to 37 (for TS10000). Because, in practice, we do not expect very high densities of group member nodes relative to the Internet topology itself, worst-case stress using CAN-based multicast should be at a reasonable level. In future work, we intend looking into techniques that might further lower this stress value.

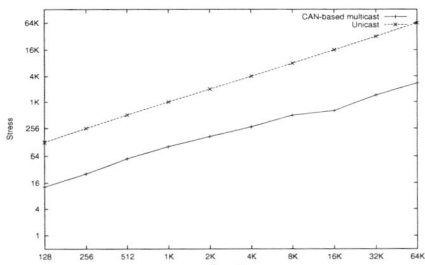 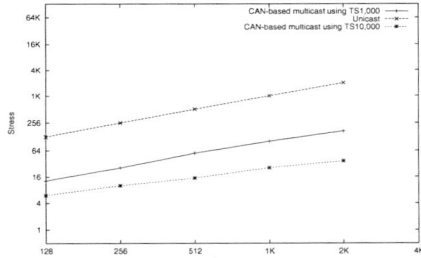

Fig. 12. *Stress   versus   Increasing Group Size.*

Fig. 13. *Effect of Topology Density on Stress.*

5   Related Work

The case for application-level multicast as a more tractable alternative to a network-level multicast service was first put forth in [4,3,1].

The End-system multicast [4] work proposes an architecture for multicast over small and sparse groups. End-system multicast builds a mesh structure across participating end-hosts and then constructs source-rooted trees by running a routing protocol over this mesh. The authors also study the fundamental performance penalty associated with such an application-level model. The authors in [1] argue for infrastructure support to tackle the problem of content distribution over the Internet. The Scattercast architecture relies on proxies deployed within the network infrastructure. These proxies self-organise into an application-level mesh over which a global routing algorithm is used to construct distribution trees. In terms of being a solution to application-level multicast, the key difference between our work and the End-System multicast and Scattercast work is the potential for CAN-based multicast to scale to large group sizes.

Yoid [3] proposes a solution to application-level multicast wherein a spanning tree is directly constructed across the participating nodes without first constructing a mesh structure. The resultant protocols are more complex because the tree-first approach results in expensive loop detection and avoidance techniques and must be made resilient to partitions.

Tapestry [10] is a wide-area overlay routing and location infrastructure that, like CANs, embeds nodes in a well-defined virtual address space. Bayeux [11] is a source-specific, application-level multicast scheme that leverages the Tapestry routing infrastructure. To join a multicast session, Bayeux nodes send *JOIN* messages to the source node. The source replies to a *JOIN* request by routing a *TREE* message, on the Tapestry overlay, to the requesting node. This *TREE* message is used to set up state at intermediate nodes along the path from the source node to the new member. Similarly, a *LEAVE* message from an existing member triggers a *PRUNE* message from the root, which removes the appropriate forwarding state along the distribution tree. Bayeux and CAN-based multicast are similar in that they achieve scalability by leveraging the scalable routing infrastructure provided by systems like CAN and Tapestry. In terms of service model, Bayeux fundamentally supports only source-specific multicast while CAN-based multicast allows any group member to act as a traffic source. In terms of design, Bayeux uses an explicit protocol to set-up and tear down a distribution tree from the source node to the current set of receiver nodes. CAN-based multicast by contrast, fully exploits the CAN structure because of which messages can be forwarded without requiring a routing protocol to explicitly construct distribution trees.

Overcast[5] is a scheme for source-specific, reliable multicast using an overlay network. Overcast constructs efficient dissemination trees rooted at the single source of traffic. The overlay network in Overcast is composed of nodes that reside within the network infrastructure. This assumption of the existence of permanent storage within the network distinguishes Overcast from CANs and indeed, from most of the other systems described above. Unlike Overcast, CANs can be composed entirely from end-user machines with no form of central authority.

6   Conclusion

Content-Addressable Networks have the potential to serve as an infrastructure that is useful across a range of applications. In this paper, we present and evaluate a scheme that extends the basic CAN framework to support application-level multicast delivery. There are, we believe, two key benefits to CAN-based multicast: the potential to scale to large groups without restricting the service model and the simplicity of the scheme under the assumption of the deployment of a distributed infrastructure such as a Content-Addressable Network.

Our CAN-based multicast scheme is optimal in terms of the distance (in terms of path length) in flooding messages over the CAN overlay structure itself. In future work, we intend looking into simple clustering techniques to further reduce the link stress caused by our flooding algorithm and understanding what the fundamental limitations there are. A number of important questions such as security, loss recovery, and congestion control remain to be addressed in the context of CAN-based multicast.

Acknowledgements

We thank Ion Stoica for his valuable input and Yan Chen and Morley Mao for sharing their data.

References

1. Y. Chawathe, S. McCanne, and E. Brewer. An architecture for internet content distribution as an infrastructure service. Available at
   `http://www.cs.berkeley.edu/~yatin/papers`, 2000.
2. S. E. Deering. *Multicast Routing in a Datagram Internetwork*. PhD thesis, Stanford University, Dec. 1991.
3. P. Francis. Yoid: Extending the internet multicast architecture. Unpublished paper, available at
   `http://www.aciri.org/yoid/docs/index.html`, Apr. 2000.
4. Y. hua Chu, S. Rao, and H. Zhang. A case for end system multicast. In *Proceedings of SIGMETRICS 2000*, Santa Clara, CA, June 2000.
5. J. Jannotti, D. Gifford, K. Johnson, F. Kaashoek, and J. O'Toole. Overcast: Reliable multicasting with an overlay network. In *Proceedings of the Fourth Symposium on Operating Systems Design and Implementation*, San Diego, CA, Oct. 2000.
6. S. Ratnasamy, P. Francis, M. Handley, R. Karp, and S. Shenker. A Scalable Content-Addressable Network. In *Proceedings of SIGCOMM 2001*, Aug. 2001.
7. A. Rowstron and P. Druschel. Pastry: Scalable, distributed object location and routing for large-scale peer-to-peer systems. Available at
   `http://research.microsoft.com/~antr/PAST/`, 2001.
8. I. Stoica, R. Morris, D. Karger, M. F. Kaashoek, and H. Balakrishnan. Chord: A scalable peer-to-peer lookup service for internet applications. In *Proceedings of SIGCOMM 2001*, Aug. 2001.
9. E. Zegura, K. Calvert, and S. Bhattacharjee. How to Model an Internetwork. In *Proceedings IEEE Infocom '96*, San Francisco, CA, May 1996.
10. B. Y. Zhao, J. Kubiatowicz, and A. Joseph. Tapestry: An infrastructure for fault-tolerant wide-area location and routing. Available at
    `http://www.cs.berkeley.edu/~ravenben/tapestry/`, 2001.
11. S. Q. Zhuang, B. Zhao, A. Joseph, R. Katz, and J. Kubiatowicz. Bayeux: An architecture for scalable and fault-tolerant wide-area data dissemination. In *Proceedings of the Eleventh International Workshop on Network and OS Support for Digital Audio and Video*, New York, July 2001. ACM.

SCRIBE: The Design of a Large-Scale Event Notification Infrastructure

Antony Rowstron[1], Anne-Marie Kermarrec[1],
Miguel Castro[1], and Peter Druschel[2]

[1] Microsoft Research
7 J J Thomson Avenue, Cambridge, CB3 0FB, UK
{antr,anne-mk,mcastro}@microsoft.com
[2] Rice University MS-132, 6100 Main Street
Houston, TX 77005-1892, USA
druschel@cs.rice.edu

Abstract. This paper presents Scribe, a large-scale event notification infrastructure for topic-based publish-subscribe applications. Scribe supports large numbers of topics, with a potentially large number of subscribers per topic. Scribe is built on top of Pastry, a generic peer-to-peer object location and routing substrate overlayed on the Internet, and leverages Pastry's reliability, self-organization and locality properties. Pastry is used to create a topic (group) and to build an efficient multicast tree for the dissemination of events to the topic's subscribers (members). Scribe provides weak reliability guarantees, but we outline how an application can extend Scribe to provide stronger ones.

1 Introduction

Publish-subscribe has emerged as a promising paradigm for large-scale, Internet based distributed systems. In general, subscribers register their interest in a topic or a pattern of events and then asynchronously receive events matching their interest, regardless of the events' publisher. Topic-based publish-subscribe [1,2,3] is very similar to group-based communication; subscribing is equivalent to becoming a member of a group. For such systems the challenge remains to build an infrastructure that can scale to, and tolerate the failure modes of the general Internet.

Techniques such as SRM (Scalable Reliable Multicast Protocol) [4] or RMTP (Reliable Message Transport Protocol) [5] have added reliability to network-level IP multicast [6,7] solutions. However, tracking membership remains an issue in router-based multicast approaches and the lack of wide deployment of IP multicast limits their applicability. As a result, application-level multicast is gaining popularity. Appropriate algorithms and systems for scalable subscription management and scalable, reliable propagation of events are still an active research area [8,9,10,11].

Recent work on peer-to-peer overlay networks offers a scalable, self-organizing, fault-tolerant substrate for decentralized distributed applications [12,13,14,15].

J. Crowcroft and M. Hofmann (Eds.): NGC 2001, LNCS 2233, pp. 30–43, 2001.

Such systems offer an attractive platform for publish-subscribe systems that can leverage these properties. In this paper we present Scribe, a large-scale, decentralized event notification infrastructure built upon Pastry, a scalable, self-organizing peer-to-peer location and routing substrate with good locality properties [12]. Scribe provides efficient application-level multicast and is capable of scaling to a large number of subscribers, publishers and topics.

Scribe and Pastry adopt a fully decentralized peer-to-peer model, where each participating node has equal responsibilities. Scribe builds a multicast tree, formed by joining the Pastry routes from each subscriber to a rendez-vous point associated with a topic. Subscription maintenance and publishing in Scribe leverages the robustness, self-organization, locality and reliability properties of Pastry. Section 2 gives an overview of the Pastry routing and object location infrastructure. Section 3 describes the basic design of Scribe and we discuss related work in Section 4.

2   Pastry

In this section we briefly sketch Pastry [12]. Pastry forms a secure, robust, self-organizing overlay network in the Internet. Any Internet-connected host that runs the Pastry software and has proper credentials can participate in the overlay network.

Each Pastry node has a unique, 128-bit nodeId. The set of existing nodeIds is uniformly distributed; this can be achieved, for instance, by basing the nodeId on a secure hash of the node's public key or IP address. Given a message and a key, Pastry reliably routes the message to the Pastry node with a nodeId that is numerically closest to the key, among all live Pastry nodes. Assuming a Pastry network consisting of N nodes, Pastry can route to any node in less than $\lceil log_{2^b} N \rceil$ steps on average (b is a configuration parameter with typical value 4). With concurrent node failures, eventual delivery is guaranteed unless $\lfloor l/2 \rfloor$ nodes with *adjacent* nodeIds fail simultaneously (l is a configuration parameter with typical value 16).

The tables required in each Pastry node have only $(2^b - 1) * \lceil log_{2^b} N \rceil + 2l$ entries, where each entry maps a nodeId to the associated node's IP address. Moreover, after a node failure or the arrival of a new node, the invariants in all affected routing tables can be restored by exchanging $O(log_{2^b} N)$ messages. In the following, we briefly sketch the Pastry routing scheme. A full description and evaluation of Pastry can be found in [12].

For the purposes of routing, nodeIds and keys are thought of as a sequence of digits with base 2^b. A node's routing table is organized into $\lceil log_{2^b} N \rceil$ rows with $2^b - 1$ entries each. The $2^b - 1$ entries in row n of the routing table each refer to a node whose nodeId matches the present node's nodeId in the first n digits, but whose $n + 1$th digit has one of the $2^b - 1$ possible values other than the $n + 1$th digit in the present node's id. The uniform distribution of nodeIds ensures an even population of the nodeId space; thus, only $\lceil log_{2^b} N \rceil$ levels are populated in the routing table. Each entry in the routing table refers to one of potentially

NodeId 10233102

Leaf set

SMALLER		LARGER	
10233033	10233021	10233120	10233122
10233001	10233000	10233230	10233232

Routing table

-0-2212102	1	-2-2301203	-3-1203203
0	1-1-301233	1-2-230203	1-3-021022
10-0-31203	10-1-32102	2	10-3-23302
102-0-0230	102-1-1302	102-2-2302	3
1023-0-322	1023-1-000	1023-2-121	3
10233-0-01	1	10233-2-32	
0		102331-2-0	
		2	

Neighborhood set

13021022	10200230	11301233	31301233
02212102	22301203	31203203	33213321

Fig. 1. State of a hypothetical Pastry node with nodeId 10233102, $b = 2$. All numbers are in base 4. The top row of the routing table represents level zero. The neighborhood set is not used in routing, but is needed during node addition/recovery.

many nodes whose nodeId have the appropriate prefix. Among such nodes, the one closest to the present node (according to a scalar proximity metric, such as the delay or the number of IP routing hops) is chosen in practice.

In addition to the routing table, each node maintains IP addresses for the nodes in its *leaf set*, i.e., the set of nodes with the $l/2$ numerically closest larger nodeIds, and the $l/2$ nodes with numerically closest smaller nodeIds, relative to the present node's nodeId. Figure 1 depicts the state of a hypothetical Pastry node with the nodeId 10233102 (base 4), in a system that uses 16 bit nodeIds and a value of $b = 2$.

In each routing step, a node normally forwards the message to a node whose nodeId shares with the key a prefix that is at least one digit (or b bits) longer than the prefix that the key shares with the present node's id. If no such node is found in the routing table, the message is forwarded to a node whose nodeId shares a prefix with the key as long as the current node, but is numerically closer to the key than the present node's id. Such a node must be in the leaf set unless the message has already arrived at the node with numerically closest nodeId or its neighbor. And, unless $\lfloor |l|/2 \rfloor$ adjacent nodes in the leaf set have failed simultaneously, at least one of those nodes must be live.

2.1 Locality

Next, we discuss Pastry's locality properties, i.e., the properties of Pastry's routes with respect to the proximity metric. The proximity metric is a scalar value that reflects the "distance" between any pair of nodes, such as the number of IP routing hops, geographic distance, delay, or a combination thereof. It is assumed that a function exists that allows each Pastry node to determine the "distance" between itself and a node with a given IP address.

We limit our discussion to two of Pastry's locality properties that are relevant to Scribe. The first property is the total distance, in terms of the proximity metric, that messages are traveling along Pastry routes. Recall that each entry in the node routing tables is chosen to refer to the nearest node, according to the proximity metric, with the appropriate nodeId prefix. As a result, in each step a message is routed to the nearest node with a longer prefix match. Simulations show that, given a network topology based on the Georgia Tech model [16], the average distance traveled by a message is less than 66% higher than the distance between the source and destination in the underlying Internet.

Let us assume that two nodes within distance d from each other route messages with the same key, such that the distance from each node to the node with nodeId closest to the key is much larger than d. The second locality property is concerned with the "distance" the messages travel until they reach a node where their routes merge. Simulations show that the average distance traveled by each of the two messages before their routes merge is approximately equal to the distance between their respective source nodes. These properties have a strong impact on the locality properties of the Scribe multicast trees, as explained in Section 3.

2.2   Node Addition and Failure

A key design issue in Pastry is how to efficiently and dynamically maintain the node state, i.e., the routing table, leaf set and neighborhood sets, in the presence of node failures, node recoveries, and new node arrivals. The protocol is described and evaluated in [12].

Briefly, an arriving node with the newly chosen nodeId X can initialize its state by contacting a nearby node A (according to the proximity metric) and asking A to route a special message using X as the key. This message is routed to the existing node Z with nodeId numerically closest to X. X then obtains the leaf set from Z, the neighborhood set from A, and the ith row of the routing table from the ith node encountered along the route from A to Z. One can show that using this information, X can correctly initialize its state and notify nodes that need to know of its arrival, thereby restoring all of Pastry's invariants.

To handle node failures, neighboring nodes in the nodeId space (which are aware of each other by virtue of being in each other's leaf set) periodically exchange keep-alive messages. If a node is unresponsive for a period T, it is presumed failed. All members of the failed node's leaf set are then notified and they update their leaf sets to restore the invariant. Since the leaf sets of nodes with adjacent nodeIds overlap, this update is trivial. A recovering node contacts the nodes in its last known leaf set, obtains their current leaf sets, updates its own leaf set and then notifies the members of its new leaf set of its presence. Routing table entries that refer to failed nodes are repaired lazily; the details are described in [12].

2.3   Pastry API

In this section, we briefly describe the application programming interface (API) exported by Pastry which is used in the Scribe implementation. The presented API is slightly simplified for clarity. Pastry exports the following operations:

route(msg,key) causes Pastry to route the given message to the node with nodeId numerically closest to key, among all live Pastry nodes.

send(msg,IP-addr) causes Pastry to send the given message to the node with the specified IP address, if that node is live. The message is received by that node through the deliver method.

Applications layered on top of Pastry must export the following operations:

deliver(msg,key) called by Pastry when a message is received and the local node's nodeId is numerically closest to key among all live nodes, or when a message is received that was transmitted via *send*, using the IP address of the local node.

forward(msg,key,nextId) called by Pastry just before a message is forwarded to the node with nodeId = nextId. The application may change the contents of the message or the value of nextId. Setting the nextId to NULL will terminate the message at the local node.

In the following section, we will describe how Scribe is layered on top of the Pastry API. Other applications built on top of Pastry include PAST, a persistent, global storage utility [17,18].

3   Scribe

Any Scribe node may create a *topic*; other nodes can then register their interest in the topic and become a *subscriber* to the topic. Any Scribe node with the appropriate credentials for the topic can then publish events, and Scribe disseminates these events to all the topic's subscribers. Scribe provides a best-effort dissemination of events, and specifies no particular event delivery order. However, stronger reliability guarantees and ordered delivery for a topic can be built on top of Scribe, as outlined in Section 3.2. Nodes can publish events, create and subscribe to many topics, and topics can have many publishers and subscribers. Scribe can support large numbers of topics with a wide range of subscribers per topic, and a high rate of subscriber turnover.

Scribe offers a simple API to its applications:

create(credentials, topicId) creates a topic with topicId. Throughout, the credentials are used for access control.

subscribe(credentials, topicId, eventHandler) causes the local node to subscribe to the topic with topicId. All subsequently received events for that topic are passed to the specified event handler.

unsubscribe(credentials, topicId) causes the local node to unsubscribe from the topic with topicId.

publish(credentials, topicId, event) causes the event to be published in the topic with topicId.

Scribe uses Pastry to manage topic creation, subscription, and to build a per-topic multicast tree used to disseminate the events published in the topic. Pastry and Scribe are fully decentralized, all decisions are based on local information, and each node has identical capabilities. Each node can act as a publisher, a root of a multicast tree, a subscriber to a topic, a node within a multicast tree, and any sensible combination of the above. Much of the scalability and reliability of Scribe and Pastry derives from this peer-to-peer model.

3.1 Scribe Implementation

A Scribe system consists of a network of Pastry nodes, where each node runs the Scribe application software. The Scribe software on each node provides the *forward* and *deliver* methods, which are invoked by Pastry whenever a Scribe message arrives. The pseudo-code for these Scribe methods, simplified for clarity, is shown in Figure 2 and Figure 3, respectively.

```
(1) forward(msg, key, nextId)
(2)      switch msg.type is
(3)          SUBSCRIBE :if !(msg.topic ∈ topics)
(4)                        topics = topics ∪ msg.topic
(5)                        msg.source = thisNodeId
(6)                        route (msg,msg.topic)
(7)                        topics[msg.topic].children ∪ msg.source
(8)                        nextId = null  // Stop routing the original message
```

Fig. 2. Scribe Implementation of Forward.

```
(1) deliver(msg,key)
(2)      switch msg.type is
(3)          CREATE :       topics = topics ∪ msg.topic
(4)          SUBSCRIBE :    topics[msg.topic].children ∪ msg.source
(5)          PUBLISH :      ∀ node in topics[msg.topic].children
(6)                            send (msg,node)
(7)                         if subscribedTo(msg.topic)
(8)                            invokeEventHandler (msg.topic, msg)
(9)          UNSUBSCRIBE :  topics[msg.topic].children =
                               topics[msg.topic].children - msg.source
(10)                        if (|topics[msg.topic].children| = 0)
(11)                           msg.source = thisNodeId
(12)                           send (msg,topics[msg.topic].parent)
```

Fig. 3. Scribe Implementation of Deliver.

Recall that the forward method is called whenever a Scribe message is routed through a node. The deliver method is called when a Scribe message arrives at the node with nodeId numerically closest to the message's key, or when a message was addressed to the local node using the Pastry *send* operation. The possible message types in Scribe are SUBSCRIBE, CREATE, UNSUBSCRIBE and PUBLISH; the roles of these messages are described in the next sections.

The following variables are used in the pseudocode: *topics* is the set of topics that the local node is aware of, *msg.source* is the nodeId of the message's source node, *msg.event* is the published event (if present), *msg.topic* is the topicId of the topic and *msg.type* is the message type.

Topic Management. Each topic has a unique *topicId*. The Scribe node with a nodeId numerically closest to the topicId acts as the *rendez-vous point* for the associated topic. The rendez-vous point forms the root of a multicast tree created for the topic.

To create a topic, a Scribe node asks Pastry to route a CREATE message using the topicId as the key (e.g. route(CREATE,topicId)). Pastry delivers this message to the node with the nodeId numerically closest to topicId. The Scribe deliver method adds the topic to the list of topics it already knows about (line 3 of Figure 3). It also checks the credentials to ensure that the topic can be created, and stores the credentials in the topics set. This Scribe node becomes the rendez-vous point for the topic.

The topicId is the hash of the topic's textual name concatenated with its creator's name. The hash is computed using a collision resistant hash function (e.g. SHA-1 [19]), which ensures a uniform distribution of topicIds. Since Pastry nodeIds are also uniformly distributed, this ensures an even distribution of topics across Pastry nodes. A topicId can be generated by any Scribe node using only the textual name of the topic and its creator, without the need for an additional naming service. Of course, proper credentials are necessary to subscribe or publish in the associated topic.

Membership Management. Scribe creates a multicast tree, rooted at the rendez-vous point, to disseminate the events published in the topic. The multicast tree is created using a scheme similar to reverse path forwarding [20]. The tree is formed by joining the Pastry routes from each subscriber to the rendez-vous point. Subscriptions to a topic are managed in a decentralized manner to support large and dynamic sets of subscribers.

Scribe nodes that are part of a topic's multicast tree are called *forwarders* with respect to the topic; they may or may not be subscribers to the topic. Each forwarder maintains a *children table* for the topic containing an entry (IP address and NodeId) for each of its children in the multicast tree.

When a Scribe node wishes to subscribe to a topic, it asks Pastry to route a SUBSCRIBE message with the topic's *topicId* as the key (e.g. route (SUBSCRIBE,topicId)). This message is routed by Pastry towards the topic's rendez-vous point. At each node along the route, Pastry invokes Scribe's forward method. Forward (lines 3 to 8 in Figure 2) checks its list of topics to see if it is currently a forwarder; if so, it accepts the node as a child, adding it to the children table. If the node is not already a forwarder, it creates an entry for the topic, and adds the source node as a child in the associated children table. It then becomes a forwarder for the topic by sending a SUBSCRIBE message to the next node along the route from the original subscriber to the rendez-vous point. The original message from the source is terminated; this is achieved by setting nextId = null, in line 8 of Figure 2.

Figure 4 illustrates the subscription mechanism. The circles represent nodes, and some of the nodes have their *nodeId* shown. For simplicity $b = 1$, so the prefix is matched one bit at a time. We assume that there is a topic with *topicId* 1100 whose rendez-vous point is the node with the same identifier. The node with *nodeId* 0111 is subscribing to this topic. In this example, Pastry routes the SUBSCRIBE message to node 1001; then the message from 1001 is routed to 1101; finally, the message from 1101 arrives at 1100. This route is indicated by the solid arrows in Figure 4.

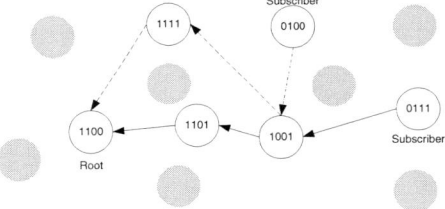

Fig. 4. Base Mechanism for Subscription and Multicast Tree Creation.

Let us assume that nodes 1001 and 1101 are not already forwarders for topic 1100. The subscription of node 0111 causes the other two nodes along the route to become forwarders for the topic, and causes them to add the preceding node in the route to their children tables. Now let us assume that node 0100 decides to subscribe to the same topic. The route that its SUBSCRIBE message would take is shown using dot-dash arrows. Since node 1001 is already a forwarder, it adds node 0100 to its children table for the topic, and the SUBSCRIBE message is terminated.

When a Scribe node wishes to unsubscribe from a topic, a node locally marks the topic as no longer required. If there are no entries in the children table, it sends a UNSUBSCRIPTION message to its parent in the multicast tree, as shown in lines 9 to 12 in Figure 3. The message proceeds recursively up the multicast tree, until a node is reached that still has entries in the children table after removing the departing child. It should be noted that nodes in the multicast tree are aware of their parent's nodeId only after they have received an event from their parent. Should a node wish to unsubscribe before receiving an event, the implementation transparently delays the unsubscription until the first event is received.

The subscriber management mechanism is efficient for topics with different numbers of subscribers, varying from one to all Scribe nodes. The list of subscribers to a topic is distributed across the nodes in the multicast tree. Pastry's randomization properties ensure that the tree is well balanced and that the forwarding load is evenly balanced across the nodes. This balance enables

Scribe to support large numbers of topics and subscribers per topics. Subscription requests are handled locally in a decentralized fashion. In particular, the rendez-vous point does not handle all subscription requests.

The locality properties of Pastry (discussed in Section 2.1) ensure that the network routes from the root to each subscriber are short with respect to the proximity metric. In addition, subscribers that are close with respect to the proximity metric tend to be children of a parent in the multicast tree that is also close to them. This reduces stress on network links because the parent receives a single copy of the event message and forwards copies to its children along short routes.

Event Dissemination. Publishers use Pastry to locate the rendez-vous point of a topic. If the publisher is aware of the rendez-vous point's IP address then the PUBLISH message can be sent straight to the node. If the publisher does not know the IP address of the rendez-vous point, then it uses Pastry to route to that node (e.g. route(PUBLISH, topicId)), and asks the rendez-vous point to return its IP address to the publisher. Events are disseminated from the rendez-vous point along the multicast tree in the obvious way (lines 5 and 6 of Figure 3).

The caching of the rendez-vous point's IP address is an optimization, to avoid repeated routing through Pastry. If the rendez-vous point fails then the publisher can route the event through Pastry and discover the new rendez-vous point. If the rendez-vous point has changed because a new node has arrived, then the old rendez-vous point can forward the publish message to the new rendez-vous point and ask the new rendez-vous point to forward its IP address to the publisher.

There is a single multicast tree for each topic and all publishers use the above procedure to publish events. This allows the rendez-vous node to perform access control.

3.2   Reliability

Publish/subscribe applications may have diverse reliability requirements. Some topics may require reliable and ordered delivery of events, whilst others require only best-effort delivery. Therefore, Scribe provides only best-effort delivery of events but it offers a framework for applications to implement stronger reliability guarantees.

Scribe uses TCP to disseminate events reliably from parents to their children in the multicast tree, and it uses Pastry to repair the multicast tree when a forwarder fails.

Repairing the Multicast Tree. Periodically, each non-leaf node in the tree sends a heartbeat message to its children. When events are frequently published on a topic, most of these messages can be avoided since events serve as an implicit heartbeat signal. A child suspects that its parent is faulty when it fails to receive heartbeat messages. Upon detection of the failure of its parent, a node calls Pastry to route a SUBSCRIBE message to the topic's identifier. Pastry will route the message to a new parent, thus repairing the multicast tree.

For example, in Figure 4, consider the failure of node 1101. Node 1001 detects the failure of 1101 and uses Pastry to route a SUBSCRIBE message towards the root through an alternative route. The message reaches node 1111, which adds 1001 to its children table and, since it is not a forwarder, sends a SUBSCRIBE message towards the root. This causes node 1100 to add 1111 to its children table.

Scribe can also tolerate the failure of multicast tree roots (rendez-vous points). The state associated with the rendez-vous point, which identifies the topic creator and has an access control list, is replicated across the k closest nodes to the root node in the nodeId space (where a typical value of k is 5). It should be noted that these nodes are in the leaf set of the root node. If the root fails, its immediate children detect the failure and subscribe again through Pastry. Pastry routes the subscriptions to a new root (the live node with the numerically closest nodeId to the topicId), which takes over the role of the rendez-vous point. Publishers likewise discover the new rendez-vous point by routing via Pastry.

Children table entries are discarded unless they are periodically refreshed by an explicit message from the child, stating its continued interest in the topic.

This tree repair mechanism scales well: fault detection is done by sending messages to a small number of nodes, and recovery from faults is local; only a small number of nodes ($O(log_{2^b} N)$) is involved.

Providing Additional Guarantees. By default, Scribe provides reliable, ordered delivery of events only if the TCP connections between the nodes in the multicast tree do not break. For example, if some nodes in the multicast tree fail, Scribe may fail to deliver events or may deliver them out of order.

Scribe provides a simple mechanism to allow applications to implement stronger reliability guarantees. Applications can define the following upcall methods, which are invoked by Scribe.

forwardHandler(msg) is invoked by Scribe before the node forwards an event, msg, to its children in the multicast tree. The method can modify msg before it is forwarded.

subscribeHandler(msg) is invoked by Scribe after a new child is added to one of the node's children tables. The argument is the SUBSCRIBE message.

faultHandler(msg) is invoked by Scribe when a node suspects that its parent is faulty. The argument is the SUBSCRIBE message that is sent to repair the tree. The method can modify msg to add additional information before it is sent.

For example, an application can implement ordered, reliable delivery of events by defining the upcalls as follows. The forwardHandler is defined such that the root assigns a sequence number to each event and such that recently published events are buffered by the root and by each node in the multicast tree. Events are retransmitted after the multicast tree is repaired. The faultHandler adds the last sequence number, n, delivered by the node to the SUBSCRIBE message and the subscribeHandler retransmits buffered events with sequence numbers above n to the new child. To ensure reliable delivery, the events must be buffered for an amount of time that exceeds the maximal time to repair the multicast tree after a TCP connection breaks.

To tolerate root failures, the root needs to be replicated. For example, one could choose a set of replicas in the leaf set of the root and use an algorithm like Paxos [21] to ensure strong consistency.

4   Related Work

Like Scribe, Overcast [22] and Narada [23] implement multicast using a self-organizing overlay network, and they assume only unicast support from the underlying network layer. Overcast builds a source-rooted multicast tree using end-to-end bandwidth measurements to optimize bandwidth between the source and the various group members. Narada uses a two step process to build the multicast tree. First, it builds a mesh per group containing all the group members. Then, it constructs a spanning tree of the mesh for each source to multicast data. The mesh is dynamically optimized by performing end-to-end latency measurements and adding and removing links to reduce multicast latency. The mesh creation and maintenance algorithms assume that all group members know about each other and, therefore, do not scale to large groups.

Scribe builds a multicast tree on top of a Pastry network, and relies on Pastry to optimize route locality based on a proximity metric (e.g. IP hops or latency). The main difference is that the Pastry network can scale to an extremely large number of nodes because the algorithms to build and maintain the network have space and time costs of $O(log_{2^b} N)$. This enables support for extremely large groups and sharing of the Pastry network by a large number of groups.

The recent work on Bayeux [11] is the most similar to Scribe. Bayeux is built on top of a scalable peer-to-peer object location system called Tapestry [13] (which is similar to Pastry). Like Scribe, it supports multiple groups, and it builds a multicast tree per group on top of Tapestry but this tree is built quite differently. Each request to join a group is routed by Tapestry all the way to the node acting as the root. Then, the root records the identity of the new member and uses Tapestry to route another message back to the new member. Every Tapestry node (or router) along this route records the identity of the new member. Requests to leave a group are handled in a similar way.

Bayeux has two scalability problems when compared to Scribe. Firstly, it requires nodes to maintain more group membership information. The root keeps a list of all group members, the routers one hop away from the route keep a list containing on average $\frac{S}{b}$ members (where b is the base used in Tapestry routing), and so on. Secondly, Bayeux generates more traffic when handling group membership changes. In particular, all group management traffic must go through the root. Bayeux proposes a multicast tree partitioning mechanism to ameliorate these problems by splitting the root into several replicas and partitioning members across them. But this only improves scalability by a small constant factor.

In Scribe, the expected amount of group membership information kept by each node is small, as the subscribers are distributed over the nodes. Additionally, group join and leave requests are handled locally. This allows Scribe to scale to extremely large groups and to deal with rapid changes in group membership efficiently.

The mechanisms for fault resilience in Bayeux and Scribe are also very different. All the mechanisms for fault resilience proposed in Bayeux are sender-based whereas Scribe uses a receiver-based mechanism. In Bayeux, routers proactively duplicate outgoing packets across several paths or perform active probes to select alternative paths. Both these schemes have some disadvantages. The mechanisms that perform packet duplication consume additional bandwidth, and the mechanisms that select alternative paths require replication and transfer of group membership information across different paths. Scribe relies on heartbeats sent by parents to their children in the multicast tree to detect faults, and children use Pastry to reroute to a different parent when a fault is detected. Additionally, Bayeux does not provide a mechanism to handle root failures whereas Scribe does.

5   Conclusions

We have presented Scribe, a large-scale and fully decentralized event notification system built on top of Pastry, a peer-to-peer object location and routing substrate overlayed on the Internet. Scribe is designed to scale to large numbers of subscribers and topics, and supports multiple publishers per topic.

Scribe leverages the scalability, locality, fault-resilience and self-organization properties of Pastry. Pastry is used to maintain topics and subscriptions, and to build efficient multicast trees. Scribe's randomized placement of topics and multicast roots balances the load among participating nodes. Furthermore, Pastry's properties enable Scribe to exploit locality to build efficient multicast trees and to handle subscriptions in a decentralized manner.

Fault-tolerance in Scribe is based on Pastry's self-organizing properties. Scribe's default reliability scheme ensures automatic adaptation of the multicast tree to node and network failures. Event dissemination is performed on a best-effort basis; consistent ordering of delivered events is not guaranteed. However, stronger reliability models can be layered on top of Scribe.

Simulation results, based on a realistic network topology model and presented in [24], indicate that Scribe scales well. It efficiently supports a large number of nodes, topics, and a wide range of subscribers per topic. Hence, Scribe can concurrently support applications with widely different characteristics. Results also show that it balances the load among participating nodes, while achieving acceptable delay and link stress, when compared to network-level (IP) multicast.

References

1. Talarian Corporation. *Everything You need to know about Middleware: Mission-Critical Interprocess Communication (White Paper).* http://www.talarian.com/, 1999.
2. TIBCO. *TIB/Rendezvous White Paper.* http://www.rv.tibco.com/whitepaper.html, 1999.
3. P.T. Eugster, P. Felber, R. Guerraoui, and A.-M. Kermarrec. The many faces of publish/subscribe. Technical Report DSC ID:2000104, EPFL, January 2001.
4. S. Floyd, V. Jacobson, C.G. liu, S. McCanne, and L. Zhang. A reliable multicast framework for light-weight sessions and application level framing. *IEEE/ACM Transaction on networking*, pages 784–803, December 1997.
5. J.C. Lin and S. Paul. A reliable multicast transport protocol. In *Proc. of IEEE INFOCOM'96*, pages 1414–1424, 1996.
6. S. Deering and D. Cheriton. Multicast Routing in Datagram Internetworks and Extended LANs. *ACM Transactions on Computer Systems*, 8(2), May 1990.
7. S. Deering, D. Estrin, D. Farinacci, V. Jacobson, C. Liu, and L. Wei. The PIM Architecture for Wide-Area Multicast Routing. *IEEE/ACM Transactions on Networking*, 4(2), April 1996.
8. K.P. Birman, M. Hayden, O.Ozkasap, Z. Xiao, M. Budiu, and Y. Minsky. Bimodal multicast. *ACM Transactions on Computer Systems*, 17(2):41–88, May 1999.
9. Patrick Eugster, Sidath Handurukande, Rachid Guerraoui, Anne-Marie Kermarrec, and Petr Kouznetsov. Lightweight probabilistic broadcast. In *Proceedings of The International Conference on Dependable Systems and Networks (DSN 2001)*, July 2001.
10. Luis F. Cabrera, Michael B. Jones, and Marvin Theimer. Herald: Achieving a global event notification service. In *HotOS VIII*, May 2001.
11. Shelly Q. Zhuang, Ben Y. Zhao, Anthony D. Joseph, Randy H. Katz, and John Kubiatowicz. Bayeux: An Architecture for Scalable and Fault-tolerant Wide-Area Data Dissemination. In *Proc. of the Eleventh International Workshop on Network and Operating System Support for Digital Audio and Video (NOSSDAV 2001)*, June 2001.
12. Antony Rowstron and Peter Druschel. Pastry: Scalable, distributed object location and routing for large-scale peer-to-peer systems. In *Proc. IFIP/ACM Middleware 2001*, Heidelberg, Germany, November 2001.
13. Ben Y. Zhao, John D. Kubiatowicz, and Anthony D. Joseph. Tapestry: An infrastructure for fault-resilient wide-area location and routing. Technical Report UCB//CSD-01-1141, U. C. Berkeley, April 2001.
14. I. Stoica, R. Morris, D. Karger, M. F. Kaashoek, and H. Balakrishnan. Chord: A scalable peer-to-peer lookup service for Internet applications. In *Proc. ACM SIGCOMM 2001*, San Diego, CA, August 2001.
15. S. Ratnasamy, P. Francis, M. Handley, R. Karp, and S. Shenker. A Scalable Content-Addressable Network. In *Proc. of ACM SIGCOMM*, August 2001.
16. E. Zegura, K. Calvert, and S. Bhattacharjee. How to model an internetwork. In *INFOCOM96*, 1996.
17. Peter Druschel and Antony Rowstron. PAST: A persistent and anonymous store. In *HotOS VIII*, May 2001.
18. Antony Rowstron and Peter Druschel. Storage management and caching in PAST, a large-scale, persistent peer-to-peer storage utility. In *Proc. ACM SOSP 2001*, Banff, Canada, October 2001.

19. FIPS 180-1. Secure hash standard. Technical Report Publication 180-1, Federal Information Processing Standard (FIPS), National Institute of Standards and Technology, US Department of Commerce, Washington D.C., April 1995.
20. Yogen K. Dalal and Robert Metcalfe. Reverse path forwarding of broadcast packets. *Communications of the ACM*, 21(12):1040–1048, 1978.
21. L. Lamport. The Part-Time Parliament. Report Research Report 49, Digital Equipment Corporation Systems Research Center, Palo Alto, CA, September 1989.
22. John Jannotti, David K. Gifford, Kirk L. Johnson, M. Frans Kaashoek, and James W. O'Toole. Overcast: Reliable Multicasting with an Overlay Network. In *Proc. of the Fourth Symposium on Operating System Design and Implementation (OSDI)*, pages 197–212, October 2000.
23. Yang hua Chu, Sanjay G. Rao, and Hui Zhang. A case for end system multicast. In *Proc. of ACM Sigmetrics*, pages 1–12, June 2000.
24. Miguel Castro, Peter Druschel, Anne-Marie Kermarrec, and Antony Rowstron. Scribe: A large-scale and decentralized publish-subscribe infrastructure, September 2001. Submitted for publication.
    `http://www.research.microsoft.com/~antr/scribe`.

SCAMP: Peer-to-Peer Lightweight Membership Service for Large-Scale Group Communication

Ayalvadi J. Ganesh, Anne-Marie Kermarrec, and Laurent Massoulié

Microsoft Research Ltd., 7JJ Thomson Avenue, Cambridge CB3 0FB, UK
{ajg,annemk,lmassoul}@microsoft.com

Abstract. Gossip-based protocols have received considerable attention for broadcast applications due to their attractive scalability and reliability properties. The reliability of probabilistic gossip schemes studied so far depends on each user having knowledge of the global membership and choosing gossip targets uniformly at random. The requirement of global knowledge is undesirable in large-scale distributed systems.
In this paper, we present a novel peer-to-peer membership service which operates in a completely decentralized manner in that nobody has global knowledge of membership. However, membership information is replicated robustly enough to support gossip with high reliability. Our scheme is completely self-organizing in the sense that the size of local views naturally converges to the 'right' value for gossip to succeed. This 'right' value is a function of system size, but is achieved without any node having to know the system size. We present the design, theoretical analysis and preliminary evaluation of SCAMP. Simulations show that its performance is comparable to that of previous schemes which use global knowledge of membership at each node.
Keywords: Scalability, reliability, peer-to-peer, gossip-based probabilistic multicast, membership, group communication, random graphs.

1 Introduction

The demand for large-scale event dissemination in distributed systems is growing rapidly but traditional network-level protocols and broadcast algorithms do not scale to more than thousands of participants [9]. Techniques such as SRM (Scalable Reliable Multicast Protocol) [6] or RMTP (Reliable Message Transport Protocol) [9] have added reliability to network-level IP multicast [3,4] solutions, using acknowledgments and repair mechanisms. However, no feature is available for membership tracking in network-level multicast approaches and their applicability is limited by the lack of wide deployment of IP multicast. As a result, application-level multicast, and in particular, gossip-based broadcast algorithms, have recently emerged as an attractive alternative. Probabilistic versions of these have received much attention and provide good scalability and reliability properties [2,10,12]. Their scalability relies on a peer-to-peer interaction model, where each participating node is in charge of a part of the dissemination process: the first time a node receives each notification, it forwards it to a random subset of other nodes (see the next Section for details). The protocols incorporate redundant messages which make them highly resilient to failures.

J. Crowcroft and M. Hofmann (Eds.): NGC 2001, LNCS 2233, pp. 44–55, 2001.

Though the above gossip-based approaches have proven scalable, they rely on a non-scalable membership protocol: they assume that the subset of nodes is chosen uniformly among all participating nodes, requiring that each node should know every other node. This imposes high requirements on memory and synchronization, which adversely affects their scalability. This has motivated work on distributing membership management [10,5] in order to provide each node with a partial random view of the system without any node having global knowledge of the membership.

Our understanding of scalable membership protocol should not be confused with that of [1], [7] where the aim is to provide each member of the group with an accurate and timely global view of the membership. The problem we consider is instead to provide each node with partial membership information which is sufficient to achieve reliable dissemination using a traditional gossip-based protocol. One approach to this issue is presented in [11], where a connection graph called a Harary graph is constructed. Optimality properties of Harary graphs ensure a good trade-off between the number of messages propagated and the reliability guarantees. However, building such a graph requires global knowledge of membership, and maintaining such a graph structure in the presence of arrivals/departures of nodes might prove difficult.

A protocol that does not rely on global knowledge of membership is *Lpbcast* [5]. However, the size of the partial view and the number of gossip targets are fixed *a priori*, which precludes decentralized adaptation to changes in system size.

We seek to provide a fully decentralized membership scheme, which meets the following goals: nodes obtain a partial view that adapts automatically to system size, and the view size is tuned to support gossip-based dissemination. In earlier work [8], we derived the fanout (number of gossip targets) required, as a function of system size, in order to achieve reliability. When the membership management is centralized or distributed among a few servers, the number of participants is easily determined, and the fanout can be adjusted to match reliability requirements. However, in a fully decentralized model, where each node operates with an incomplete view of the system, this is no longer straightforward.

We propose a novel probabilistic scalable membership protocol (SCAMP) aimed at addressing this problem. SCAMP is very simple, fully decentralized and self-configuring. As the number of participating nodes, n, increases, we show both analytically and through simulation that the size of local views automatically adapts to the desired value of $(c + 1) \log n$. Here, c is a design parameter which specifies the degree of robustness to failures: it follows from [8, Theorem 1] that any proportion of failed links up to $c/(c + 1)$ can be tolerated when the fanout is set to $(c + 1) \log n$. Preliminary evaluation results show that gossip based on the partial views provided by SCAMP is as resilient to failures as gossip based on random choice from a global membership known at each node. SCAMP can potentially be incorporated in existing gossip-based schemes to reduce memory and synchronization overhead due to membership management.

The remainder of the paper is organized as follows. We describe SCAMP in Section 2. The theoretical analysis is presented in Section 3 and simulation results in Section 4. We conclude in Section 5.

2   SCAMP: Peer-to-Peer Lightweight Membership Service for Large-Scale Group Communication

In this section we present the system model and the algorithms of SCAMP. The scalability of the algorithm relies on its peer to peer communication model between nodes for both membership management and gossip dissemination. We have designed SCAMP to achieve partial views of just the right size to be resilient to a given fraction of failures. This presupposes that nodes gossip to all nodes in their partial view. They could choose to gossip to a randomly chosen subset instead at the cost of reducing the fraction of failures tolerated.

In gossip-based protocols, notifications are propagated as follows. When a node generates a notification, it sends it to a random subset of other nodes. When any node receives a notification for the first time, it does the same. The question is how large this random subset should be chosen in order for all nodes to receive the notification with high probability. In earlier work [8], we proved the following result. If there are n nodes, and each node gossips to $\log n + s$ other nodes on average, then the probability that everyone gets the notification converges to $\exp(-e^{-s})$. In other words, there is a sharp threshold at $\log n$: the probability of success (everyone receiving the notification) is close to one if each node gossips to slightly more than $\log n$ nodes and close to zero if each node gossips to slightly fewer than $\log n$ nodes. We also derived expressions for how the success probability depends on the failure rate of nodes and links.

Previous work on gossip-based protocols has relied on each node having knowledge of the global membership list so that gossip targets can be chosen uniformly at random from all members. In [8], we proposed a scheme whereby a set of servers maintains the global membership list and provides individual nodes with a randomized partial view. Thus, nodes don't all need to have global information, but simply gossip to everyone in their local list. In the present work, we eliminate the need for servers and describe a fully decentralized scheme which achieves the same goals: nodes obtain a randomized partial view of the system, and the size of this view automatically scales correctly with system size, even though no node knows the system size. We now describe the details of this scheme.

2.1   Membership Management in SCAMP

Subscription. New nodes join the group by sending a subscription request to an arbitrary member. They start with a local view consisting of just the member to whom they sent their subscription request. When a node receives a new subscription request, it forwards the new node-id to all members of its own local view. It also creates c additional copies of the new subscription (c is a design parameter that determines the proportion of failures tolerated) and forwards them to randomly chosen nodes in its local view. When a node receives a forwarded subscription (2), it integrates the new subscriber in its view with a probability p which depends on the size of its view. If it doesn't keep the new subscriber, it forwards the subscription to a node randomly chosen from its local view. The system configures itself towards views of size $(c + 1) \log(n)$ on average, n being the number of nodes in the system.

Algorithm 1 depicts the pseudo-code for a node receiving a new subscription. Algorithm 2 depicts the pseudo-code for a node receiving a forwarded subscription.

---

1 Subscription Management

Upon subscription(s) of a new subscriber

{The subscription of s is forwarded to all the nodes of view}
for (i=0; i< *view.Count*; i++) do
  {For each node n in View}
  Send($view[i]$,s,forwardedSusbcription);
end for
{c additional copies of the subscription s are forwarded to random nodes of view}
for (j=0; j< c; j++) do
  randomNode=RandomChoice(view.Count);
  Send($view[randomNode]$,s,forwardedSusbcription);
end for

---

2 Handling of a Forwarded Subscription

{A node receiving a forwarded subscription adds it with the probability $p = 1/(1 + sizeOf(View))$ if it doesn't have it already}
{It forwards the subscription to a node randomly chosen in its list if it does not keep it
keep=RandomChoiceBetween0and1 ()
keep=Math.Floor((view.Count+1)*keep);
if (keep==0) and $s \notin$ view then
  view.Add(s);
else
  int i=RandomChoice(view.Count);
  n=view[i];
  send(n,s,forwardedSusbcription);
end if

Note that our membership protocol creates a distribution graph which ensures that every node is connected. This implies that, in the absence of failures or unsubscriptions, the dissemination of messages is fully reliable.

Unsubscriptions. Unsusbcriptions are handled as a gossip message and are disseminated to all members of the group. Any node that has the unsubscribing node in its partial view deletes it on receiving the unsubscription message.

Recovery from Isolation. A node becomes isolated when its identifier is present in no local views because, for example, all nodes holding its identifier have either failed or unsubscribed. Such a node has a substantial probability of remaining isolated for a long time. To overcome this problem, we propose a periodic check mechanism performed by isolated nodes. A node which has not received messages for a given period (the period is chosen to be much larger than the average time between messages[1]) will resubscribe through an arbitrary node in its partial view.

3   Analysis

We now present the theoretical analysis of the algorithm described above. We model the system as a random directed graph: nodes correspond to subscribers and there is a directed arc (x, y) whenever y is in the local list of x[2].

When a new node subscribes, the action of our algorithm is to create a random number of additional arcs, as follows. Suppose there are n members already in the group. If the new node subscribes to a node with out-degree d, then $d+c+1$ arcs are added. The new node has out-degree 1, with list consisting of just the node it subscribed to. The node receiving the subscription forwards one copy of the node-id of the subscribing node to each of its neighbors, and an additional c copies to randomly chosen neighbors. These forwarded subscriptions may be kept by the neighbors or forwarded, but are not destroyed until some node keeps them. No node keeps multiple copies of the same subscription. In practice, each node chooses whether to keep a forwarded subscription with probability inversely proportional to the length of its current list. For ease of analysis, we'll assume that new arcs are added by choosing nodes uniformly at random without replacement.

Let M_n denote the number of arcs when the number of nodes has grown to n, so that the average out-degree of each node is M_n/n. We have

$$EM_n = \left(1 + \frac{1}{n-1}\right) EM_{n-1} + c + 1,$$

from which we find that $EM_n \approx (c+1)n \log n$. If in fact $M_n = (c+1)n \log n$, and the arcs are distributed uniformly at random among the nodes, then it was shown in [8, Theorem 1] that the probability of a gossip being successful is very nearly 1 if the link failure probability is smaller than $c/(c+1)$. We shall now bound the deviation of the random quantity M_n from its mean, and show that, with high probability, M_n is very close to $(c+1)n \log n$. In other words, the proposed membership management scheme achieves the desired out-degree with

---

[1] To facilitate this, we ensure that heartbeat messages are sent if no message has been sent within this period.

[2] Note that the graph represents the logical relation of membership in local views rather than the physical topology of the underlying network. The validity of the random graph model thus relies on the way in which the membership lists are created and is not dependent on the graph structure of the physical network.

high probability, with no centralized control or even knowledge of the size of the group.

Let $\mathcal{F}_n$ denote the σ-algebra corresponding to the sequence of random graphs created after each of the first n nodes joined the group. We shall show that

$$X_n := \frac{M_n}{n} - \sum_{i=1}^{n} \frac{c+1}{i}$$

is a martingale. By the assumption that new nodes subscribe to a randomly chosen member node, we have

$$E[M_{n+1}|\mathcal{F}_n] = M_n + \frac{M_n}{n} + c + 1,$$

from which it follows that $E[X_{n+1}|\mathcal{F}_n] = X_n$, i.e., X_n is a martingale. We now estimate its variance.

Let π_n denote the empirical distribution of node out-degrees conditional on $\mathcal{F}_n$. The subscription goes to a random node whose out-degree, denoted d_n, is a random draw from π_n. Now, $d_n + c$ copies of the subscription are forwarded, and are eventually kept by nodes chosen uniformly at random (without replacement)[3]. Let $d_1, \ldots, d_{d_n+c}$ denote the out-degrees of these nodes. The new empirical distribution is

$$\pi(n+1) = \frac{n}{n+1}\pi(n) + \frac{1}{n+1}\left(\delta_1 + \sum_{k=1}^{d_n+c}(\delta_{d_k+1} - \delta_{d_k})\right), \tag{1}$$

where δ_k denotes unit mass at k. Let f_n and v_n denote the expected mean and second moment of π_n, which is a random probability distribution. Let $h_n = E[d_i d_j]$ where d_i and d_j are the out-degrees of two distinct nodes chosen uniformly at random. Let w_n denote the expected second moment of the total number of edges, M_n.

Observe that $M_{n+1} = M_n + 1 + (d_n + c)$, and so $(n+1)f_{n+1} = nf_n + 1 + f_n + c$, i.e.,

$$f_{n+1} = f_n + (c+1)/(n+1). \tag{2}$$

Moreover,

$$w_{n+1} = w_n + E[d_n^2 + 2(1+c)d_n + (1+c)^2] + 2(1+c)E[M_n] + 2E[d_n M_n]$$

$$= w_n + v_n + 2(1+c)f_n + (1+c)^2 + 2(1+c)nf_n + 2\sum_{i=1}^{n} E[d_n d_i]$$

$$= w_n + 3v_n + 2(n-1)h_n + 2(1+c)(n+1)f_n + (1+c)^2.$$

---

[3] In fact, our algorithm stores subscriptions preferentially in nodes with smaller out-degree. Ignoring this increases the variance of out-degrees and so the conclusions from the analysis presented here are expected to hold *a fortiori* for our algorithm.

We also have from (1) that

$$v_{n+1} = \frac{1}{n+1} E\left(1 + \sum_{i=1}^{d_n+c} (d_i+1)^2 + \sum_{i=d_n+c+1}^{n} d_i^2\right)$$

$$= \frac{1}{n+1}\left(1 + \sum_{i=1}^{n} E[d_i^2] + 2E\sum_{i=1}^{d_n+c} d_i + d_n + c\right)$$

$$= \frac{1}{n+1}(1 + nv_n + 2E[d_1 d_n] + 2cE[d_n] + E[d_n] + c)$$

$$= \frac{n}{n+1}v_n + \frac{2}{n+1}h_n + \frac{2c+1}{n+1}f_n + \frac{c+1}{n+1}.$$

We can eliminate h_n from the two equations above using the fact that

$$w_n = E[M_n^2] = E[(\sum_{i=1}^{n} d_i)^2] = nv_n + n(n-1)h_n,$$

from which it follows that

$$h_n = \frac{w_n - nv_n}{n(n-1)}.$$

Substituting this above and simplifying, we get the recursions:

$$w_{n+1} = \left(1 + \frac{2}{n}\right) w_n + v_n + 2(1+c)(n+1)f_n + (1+c)^2 \tag{3}$$

$$v_{n+1} = \frac{n-2}{n-1}v_n + \frac{2}{(n-1)n(n+1)}w_n + \frac{2c+1}{n+1}f_n + \frac{c+1}{n+1}. \tag{4}$$

Let $\gamma_n = w_n - n^2 f_n^2$ denote the variance of M_n, and let $\eta_n = v_n - f_n^2$ denote the expected variance of the random distribution π_n. From the above, we obtain the following recursions for γ_n and η_n:

$$\gamma_{n+1} = \left(1 + \frac{2}{n}\right)\gamma_n + \eta_n, \tag{5}$$

$$\eta_{n+1} = \frac{n-2}{n-1}\eta_n + \frac{2}{(n-1)n(n+1)}\gamma_n + \frac{1}{n+1}(f_n^2 - f_n) + \frac{c+1}{n+1} - \left(\frac{c+1}{n+1}\right)^2 \tag{6}$$

Iterating (5), we obtain the expression

$$\gamma_n = n(n+1)\sum_{k=0}^{n-1} \frac{\eta_k}{(k+1)(k+2)}. \tag{7}$$

We substitute this in (6) and use straightforward bounds to obtain the inequality

$$\eta_{n+1} \le \frac{n-2}{n-1}\eta_n + \frac{2}{n-1}\sum_{k=0}^{n-1} \frac{\eta_k}{(k+1)(k+2)} + \kappa\frac{\log^2 n}{n}, \tag{8}$$

valid for all $n \ge 2$, where κ is a suitably chosen constant (the expression for f_n entails for instance that $\kappa = 3(c+1)^2 + (c+1)/\log^2 2$ would suffice).

We now establish the following result.

Lemma 1. *There exists a constant $R > 0$ such that, for all $n \geq 2$,*

$$\eta_n \leq R \log^2 n. \tag{9}$$

Proof: Assume that we have found a constant R such that the desired inequality is satisfied for all k in the range $\{2, \ldots, n\}$. In view of (8), we obtain

$$\eta_{n+1} \leq R \log^2 n - \frac{R}{n-1} \log^2 n + \frac{2R}{n-1} \sum_{k \geq 2} \frac{\log^2 k}{(k+1)(k+2)} + \frac{\eta_0 + \eta_1}{n-1} + \kappa \frac{\log^2 n}{n}.$$

Splitting the second term into two halves, and introducing the notation

$$C = 2 \sum_{k \geq 2} \frac{\log^2 k}{(k+1)(k+2)},$$

we obtain

$$\eta_{n+1} \leq R \log^2 n + (\kappa - R/2) \frac{\log^2 n}{n} + \frac{1}{n-1} (R(C - \frac{\log^2 n}{2}) + \eta_0 + \eta_1).$$

From this last equation, we see that the induction hypothesis carries over to $n + 1$, provided the inequalities $R \geq 2\kappa$ and $R(C - \log^2 n/2) + \eta_0 + \eta_1 \leq 0$ hold. Let n_0 be the smallest index $k \geq 2$ such that $\log^2 k/2 - C > \eta_0 + \eta_1$.

We are now ready to choose the constant R. A suitable choice will be

$$R = \max \left(\max_{2 \leq k \leq n_0} \left(\eta_k / \log^2 k \right), 2\kappa, 1 \right).$$

Indeed, taking R larger than $\max_{2 \leq k \leq n_0} (\eta_k / \log^2 k)$ ensures that the induction hypothesis is satisfied in the range $k = 2, \ldots, n_0$. Taking it larger than 2κ ensures that the first inequality we need to check in order to use induction is satisfied; taking it larger than 1 ensures that, for $n \geq n_0$, the second inequality $R(c - \log^2 n) + \eta_0 + \eta_1 \leq 0$ is also satisfied, hence we can use induction from n_0 onwards.

Corollary 1. *There exists a constant $R' > 0$ such that, for all $n \geq 1$,*

$$\gamma_n \leq R' n^2. \tag{10}$$

Proof: Combining (7) and (9) yields

$$\gamma_n \leq 2n^2 \left(\eta_0 + \eta_1 + R \sum_{k \geq 2} \frac{\log^2 k}{(k+1)(k+2)} \right),$$

from which the claim of the corollary follows if we choose $R' = 2(\eta_0 + \eta_1 + RC)$, where the constant C is as in the proof of the previous lemma.

We now obtain from this corollary that $\text{Var}(X_n) = \text{Var}(M_n)/n^2 \leq R'$ for all n. As a consequence, the martingale X_n is uniformly integrable, and by the martingale convergence theorem, it converges almost surely to a finite random variable X_∞ as $n \to \infty$. In other words, the mean out-degree M_n/n is close to the target value of $(c+1) \log n$ in the precise sense that their difference converges to a finite random variable (not growing with n) as $n \to \infty$. Thus, we finally obtain from Theorem 1 of [8] that gossip in the resulting random graph reaches all participants with high probability provided the proportion of failed links is smaller than $c/(c+1)$.

4   Simulation Results

In this section we present some preliminary simulation results which confirm the theoretical analysis and show the self-organizing property of SCAMP as well as the good quality of the partial views generated. We first study the size of partial views and then provide some results comparing the resilience to failure of a gossip-based algorithm relying on SCAMP for membership management with one relying on a global scheme.

4.1   View Size

The first objective of SCAMP is to ensure that each node has a randomized partial view of the membership, of the right size to ensure successful gossip. All experiments in this section have been done with $c = 0$, i.e., the objective is to achieve an average view size of $\log(n)$. Recall that a fanout of this order is required to ensure that gossip is successful with high probability. The key result we want to confirm here is that a fully decentralized scheme as in SCAMP can provide each node with a partial view of size approximately $\log(n)$, without global membership information or synchronization between nodes.

In Figure 1, we plot the average size of partial views achieved by SCAMP against system size. The figure shows that the average list size achieved by SCAMP matches the target value very closely, supporting our claim that SCAMP is self-organizing. Figure 2 shows the distribution of list sizes of individual nodes in a 5000 node system. The distribution is unimodal with mode approximately at $\log(n)$ ($\log(5000) = 8.51$). While analytical results on the success probability of gossip were derived in [8] for two specific list size distributions, namely the deterministic and binomial distributions, we believe that the results are largely insensitive to the actual degree distribution and depend primarily on the mean degree.[4] This is corroborated by simulations.

---

[4] The claim has to be qualified somewhat as the following counterexample shows. If the fanout is n with probability $\log n/n$ and zero with probability $1 - (\log n/n)$, then the mean fanout is $\log n$ but the success probability is close to zero. Barring such extremely skewed distributions, we believe the claim to be true. An open problem is to state and prove a suitable version of this claim.

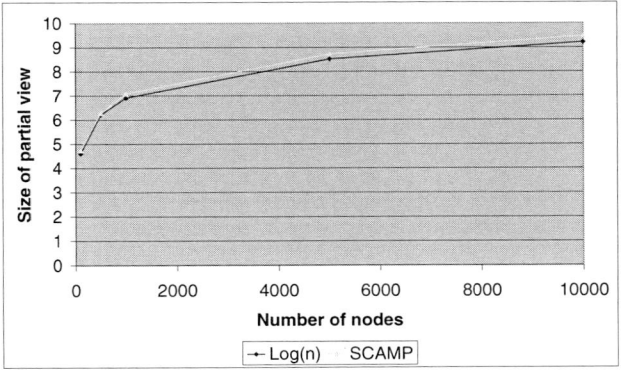

Fig. 1. Relation between System Size and Average List Size Produced by SCAMP.

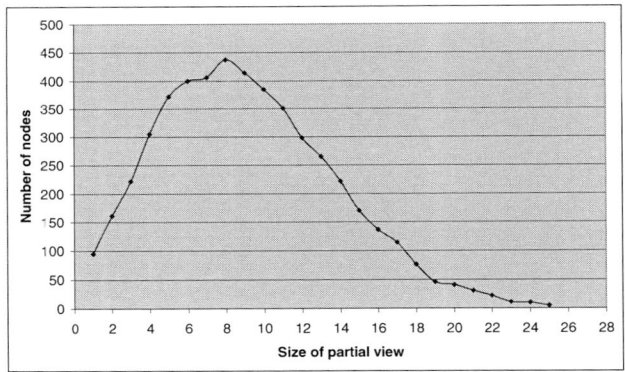

Fig. 2. Histogram of List Sizes at Individual Nodes in a 5000 Node System.

4.2   Resilience to Failures

One of the most attractive features of gossip-based multicast is its robustness to node and link failures. Event dissemination can meet stringent reliability guarantees in the presence of failures, without any explicit recovery mechanism. This makes these protocols particularly attractive in highly dynamic environments where members can disconnect for non-negligible periods and then reconnect.

We compare a gossip-based protocol relying on SCAMP with one relying on global knowledge of membership in terms of their resilience to node failures. Figure 3 depicts the simulation results. We plot the fraction of surviving nodes reached by a gossip message initiated from a random node as a function of the number of failed nodes. Two observations are notable. First, the fraction of nodes reached remains very high even when close to half the nodes have failed, which confirms the remarkable fault-tolerance of gossip-based schemes.

Second, this fraction is almost as high using SCAMP as using a scheme requiring global knowledge of membership. This attests to the quality of the partial views provided by SCAMP and demonstrates its viability as a membership scheme for supporting gossip.

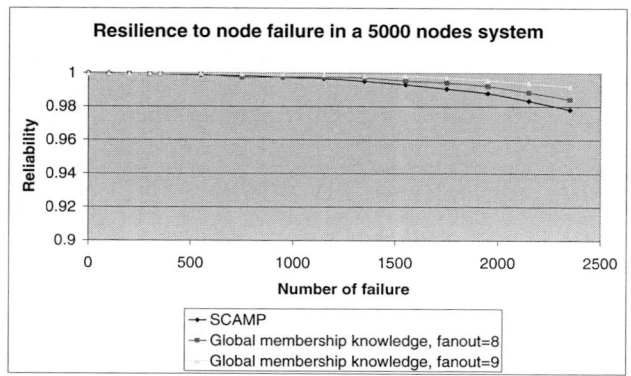

Fig. 3. Resilience to Failure in a System of 5000 Node System.

5   Conclusion

Reliable group communication is important in applications involving large-scale distributed systems. Probabilistic gossip-based protocols have proven to scale to a large number of nodes while providing attractive reliability properties. However, most gossip-based protocols rely on nodes having global membership information. For large groups, this consumes a lot of memory and generates a lot of network traffic due to the synchronization required to maintain global consistency. In order to use gossip-based algorithms in large-scale groups, which is their natural application domain, the membership protocol also needs to be decentralized and lightweight.

In this paper, we have presented the design, theoretical analysis and evaluation of SCAMP, a probabilistic peer-to-peer scalable membership protocol for gossip-based dissemination. SCAMP is fully decentralized in the sense that each node maintains only a partial view of the system. It is also self-organizing: the size of partial views naturally increases with the number of subscriptions in order to ensure the same reliability guarantees as the group grows. Thus SCAMP provides efficient support for large and highly dynamic groups.

One of the key contributions of this paper is the theoretical analysis of SCAMP, which establishes probabilistic guarantees on its performance. The analysis, which is asymptotic, is confirmed by simulations, which show that a gossip-based protocol using SCAMP as a membership service is almost as resilient to failures as a protocol relying on knowledge of global membership at each node.

Future work includes comparing SCAMP with other membership protocols, and modifying it to take geographical locality into account in the generation of partial views.

References

1. T. Anker, G.V. Chockler, D. Dolev, and I. Keidar. Scalable group membership services for novel applications. In Michael Merrit M. Mavronicolas and Nir Shavit, editors, *Networks in Distributing Computing (DIMACS workshop)*, DIMACS 45, pages 23–42. American Mathematical Society, 1998.
2. K.P. Birman, M. Hayden, O.Ozkasap, Z. Xiao, M. Budiu, and Y. Minsky. Bimodal multicast. *ACM Transactions on Computer Systems*, 17(2):41–88, May 1999.
3. S. Deering and D. Cheriton. Multicast Routing in Datagram Internetworks and Extended LANs. *ACM Transactions on Computer Systems*, 8(2), May 1990.
4. S. Deering, D. Estrin, D. Farinacci, V. Jacobson, C. Liu, and L. Wei. The PIM Architecture for Wide-Area Multicast Routing. *IEEE/ACM Transactions on Networking*, 4(2), April 1996.
5. P.T. Eugster, R. Guerraoui, S.B. Handurukande, A.-M. Kermarrec, and P. Kouznetsov. Lightweight probabilistic broadcast. In *IEEE International Conference on Dependable Systems and Networks (DSN2001*, 2001.
6. S. Floyd, V. Jacobson, C.G. liu, S. McCanne, and L. Zhang. A reliable multicast framework for light-weight sessions and application level framing. *IEEE/ACM Transaction on networking*, pages 784–803, December 1997.
7. I. Keidar, J. Sussman, K. Marzullo, and D. Dolev. A client-server oriented algorithm for virtually synchronous group memebership in wan's. In *20th International Conference on Distributed Computing Systems (ICDCS)*, pages 356–365, April 2000.
8. A.-M Kermarrec, L. Massoulié, and A.J. Ganesh. Probabilistic reliable dissemination in large-scale systems. Submitted for publication (available at http://research.microsoft.com/camdis/gossip.htm).
9. J.C. Lin and S. Paul. A reliable multicast transport protocol. In *Proc. of IEEE INFOCOM'96*, pages 1414–1424, 1996.
10. M.-J. Lin and K. Marzullo. Directional gossip: Gossip in a wide-area network. Technical Report CS1999-0622, University of California, San Diego, Computer Science and Engineering, June 1999.
11. M.-J. Lin, K. Marzullo, and S. Masini. Gossip versus deterministic flooding: Low message overhead and high-reliability for broadcasting on small networks. In *Proceedings of 14th International Symposium on DIStributed Computing (DISC 2000*, pages 253–267, Toledo, Spain, October 4-6 2000.
12. Q. Sun and D.C. Sturman. A gossip-based reliable multicast for large-scale high-throughput applications. In *Proceedings of the International conference on dependable Systems and Networks (DSN 2000)*, New York, USA, July 2000.

Extremum Feedback for Very Large Multicast Groups

Jörg Widmer[1] and Thomas Fuhrmann[2]

[1] Praktische Informatik IV, University of Mannheim, Germany
[2] The Boston Consulting Group, Munich, Germany

Abstract. In multicast communication, it is often required that feedback is received from a potentially very large group of responders while at the same time a feedback implosion needs to be prevented. To this end, a number of feedback control mechanisms have been proposed, which rely either on tree-based feedback aggregation or timer-based feedback suppression. Usually, these mechanisms assume that it is not necessary to discriminate between feedback from different receivers. However, for many applications this is not the case and feedback from receivers with certain response values is preferred (e.g., highest loss or largest delay). In this paper, we present modifications to timer-based feedback suppression mechanisms that introduce such a preference scheme to differentiate between receivers. The modifications preserve the desirable characteristic of reliably preventing a feedback implosion.

1 Introduction

Many multicast protocols require receiver feedback. For example, feedback can be used for control and identification functionality for multicast transport protocols [11] and for status reporting from receivers for congestion control [10]. In such scenarios, the size of the receiver set is potentially very large. Sessions with several million participants may be common in the future and without an appropriate feedback control mechanism a severe feedback implosion is possible.

Some multicast protocols arrange receivers in a tree hierarchy. This hierarchy can be used to aggregate receiver feedback at the inner nodes of the tree to effectively solve the feedback implosion problem. However, in many cases such a tree will not be available (e.g., for satellite links) or cannot be used for feedback aggregation (e.g., in networks without router support). For this reason, we will focus on feedback control using timer-based feedback suppression throughout the remainder of the paper.

Pure end-to-end feedback suppression mechanisms do not need any additional support except from the end-systems themselves and can thus be used for arbitrary settings. The basic mechanism of feedback suppression is to use random feedback timers at the receivers. Feedback is sent when the timer expires unless it is suppressed by a notification that another receiver (with a smaller timeout value for its feedback timer) already sent feedback.

J. Crowcroft and M. Hofmann (Eds.): NGC 2001, LNCS 2233, pp. 56–75, 2001.
© Springer-Verlag Berlin Heidelberg 2001

Most of the mechanisms presented so far assume that there is no preference as to which receivers send feedback. As we will see, for many applications this is not sufficient. Those applications require the feedback to reflect an extreme value for some parameter within the group. Multicast congestion control, for example, needs to get feedback from the receiver(s) experiencing the worst network conditions. Other examples are the polling of a large number of sensors for extreme values, online auctions where one is interested in the highest bids, and the detection of resource availability in very large distributed systems.

In this paper we propose several algorithms that favor feedback from receivers with certain characteristics while preserving the feedback implosion avoidance of the original feedback mechanism. Our algorithms can therefore be used to report extrema from very large multicast groups. In particular, they have been implemented as feedback mechanisms for the TFMCC protocol [14].

Past work related to this paper is presented in section 2. In section 3 we summarize basic properties of timer-based feedback algorithms and give some definitions to be used in our analysis. Depending on the amount of knowledge about the distribution of the values to be reported we distinguish extremum detection and feedback bias. With the former we just detect extreme values without forcing early responses from receivers with extreme values. This variant which requires no additional information about the distribution of the values is studied in section 4. With the latter we exploit knowledge about the value distribution by biasing the timers of responders. Biased feedback is studied in section 5. In both sections, we give a theoretical analysis of the properties of our feedback mechanisms and present simulations that corroborate our findings. We conclude the paper and give an outlook on future work in section 6.

2   Related Work

Feedback suppression algorithms have already been widely studied and employed. Good scalability to very large receiver sets can be achieved by exponentially distributing the receivers' feedback times. A method of round-based polling of the receiver set with exponentially increasing response probabilities was first proposed in [2] to be used as a feedback control mechanism for multicast video distribution. It was later refined by Nonnenmacher and Biersack [9], using a single feedback round with exponentially distributed random timers at the receivers. In [6], the authors compare the properties of different methods of setting the timer parameters with exponential feedback and give analytical terms and simulation results for feedback latency and response duplicates. However, none of these papers consider preferential feedback.

A simple scheme to gradually improve the values reported by the receivers is presented in [1]. Receivers continuously give feedback to control the sending rate of a multicast transmission. Since the lowest rate of the previous round is known, feedback can be limited to receivers reporting this rate or a lower rate. It is necessary to further adjust the rate limit by the largest possible increase during one round to be able to react to improved network conditions. After several rounds, the sending rate will reflect the smallest feedback value of the receiver set. While not specifically addressed in the paper, this scheme could be used in combination with exponential feedback timers for suppression within the feedback rounds to reliably prevent a feedback implosion. However, with this scheme it may still take a number of rounds to obtain the optimum feedback value.

Other algorithms not directly concerned with feedback suppression but with the detection of extremal values have been studied in the context of medium access control and resource scheduling [12,7]. The station allowed to use a shared resource is the one with the smallest contention parameter of all stations. A simple mechanism to determine this station is to use a window covering a subset of the possible contention parameters. Only stations with contention parameters within this window are allowed to respond and thus to compete for the resource. Depending on whether no, one, or several stations respond, the window boundaries are adjusted until the window only contains the minimum contention parameter. In the above papers, strategies how to optimally adjust the window with respect to available knowledge about the distribution of the contention parameters are discussed.

To our knowledge, the only work that is directly concerned with altering a non-topology based feedback suppression mechanism to solicit responses from receivers with specific metric values is presented in [3]. The authors discuss two different mechanisms, Targeted Slotting and Damping (TSD) and Targeted Iterative Probabilistic Polling (TIPP). For TSD, response values are divided into classes and the feedback mechanism is adjusted such that response times for the classes do not overlap. Responders within a better class always get to respond earlier than lower-class responders. Thus, the delay before feedback is received increases linearly with the number of empty high classes. Furthermore, it is not possible to obtain real values as feedback without the assignment of classes. To prevent implosion when many receivers fall into the same class, the response interval of a single class is divided into subintervals and the receivers are randomly spread over these intervals. It was shown in [9,6] that a uniform distribution of response times scales very poorly to large receiver sets. TIPP provides better scalability by using a polling mechanism based on the scheme presented in [2], thus having more favorable characteristics than uniform feedback timers. However, separate feedback rounds are still used for each possible feedback class. This results in very long feedback delays when the number of receivers is overestimated and the number of feedback classes is large. Underestimation will lead to a feedback implosion. As a solution, the authors propose estimating the size of the receiver set before starting the actual feedback mechanism. Determining the size of the receiver set requires one or more feedback rounds. In contrast, the mechanisms discussed in this paper only require a very rough upper bound on the number of receivers and will result in (close to) optimal feedback values within a single round. A further assumption for TSD and TIPP is that the distribution of the response values is known by the receivers. In most real scenarios this distribution is at best partially known or even completely unknown. If however the distribution is known, a feedback mechanism that guarantees optimum response values and at the same time prevents a feedback implosion can be built. Such a mechanism is presented in section 5.

3   General Considerations

Let us first summarize some previous work [2,9,6] on feedback control on which we will later base our analysis. For feedback suppression with exponentially distributed timers, each receiver gives feedback according to the following mechanism:

Algorithm 1. *(Exponential Feedback Suppression):*
Let N be an estimated upper bound on the number of potential responders[1] and T an upper bound on the amount of time by which the sending of the feedback can be delayed in order to avoid feedback implosion.

Upon receipt of a feedback request each receiver draws a random variable x uniformly distributed in $(0, 1]$ and sets its feedback timer to

$$t = T \max(0; 1 + \log_N x) \qquad (1)$$

When a receiver is notified that another receiver already gave feedback, it cancels its timer. If the feedback timer expires without the receiver having received such a notification, the receiver sends the feedback message.

Time is divided into feedback rounds, which are either implicitly or explicitly indicated to the receivers. In case continuous feedback is required, a new feedback round is started at the end of the previous one (i.e., after the first receiver gave feedback).

Extending the suggestions in [9], this algorithms sets the parameter of the exponential distribution to its optimal value $\lambda = \ln N$ and additionally introduces an offset of N^{-1} at $t = 0$ into the distribution that further improves feedback latency.

The choice of input parameters is critical for the functioning of the mechanism. While the mechanism is relatively insensitive to overestimation of the size of the receiver set, underestimation will result in a feedback implosion. Thus, a sufficiently large value for N should be chosen. Similarly, the maximum feedback delay T should be significantly larger than the network latency[2] τ among the receivers since for $T \approx \tau$ a feedback implosion is inevitable.

---

[1] The set of potential responders is formed by the participants that simultaneously want to give feedback. If no direct estimate is possible, N can be set to an upper bound on the size of the entire receiver set.

[2] With network latency we denote the average time between the sending of a feedback response by any one of the receivers and the receipt (of a notification) of this response by other receivers.

The expected delay until the first feedback is sent is

$$E[D] = \frac{T}{\ln N} \int_{1/N}^{1} \frac{(1-x)^n}{x} dx$$
$$\simeq T(1 - \log_N n) \qquad (2)$$

and the expected number of feedback messages sent is

$$E[M] = N^{\tau/T} \left(\frac{n}{N} + \left(1 - \frac{1}{N}\right)^n - \left(1 - \frac{1}{N^{\tau/T}}\right)^n \right) \qquad (3)$$

where n is the actual number of receivers. From Equation (3) we learn that $E[M]$ remains fairly constant over a large range of n (as long as $n \lesssim N$).

A derivation of Equations (2) and (3) can be found in [13] and [6] respectively.

3.1   Unicast vs. Multicast Feedback Channels

When receivers are able to multicast packets to all other receivers, feedback cancelation is immediate in that the feedback that ends the feedback round is received by other receivers at roughly the same time as by the sender.

However, the mechanism described in the previous section also works in environments where only the sender has multicast capabilities, such as in many satellite networks or networks where source-specific multicast [4] is deployed. In that case, feedback is first unicast back to the sender which then multicasts a feedback cancelation message to all receivers. This incurs an additional delay of half a round-trip time, thus roughly doubling the feedback latency of the system (in the case of symmetric transmission delays between the sender and the receivers and amongst the receivers themselves.)

In order to safeguard against loss of feedback cancelation messages with unicast feedback channels, we note that it may be necessary to let the sender send multiple cancelation messages in case multiple responses arrive at the sender and/or to repeat the previous cancelation message after a certain time interval. Loss of cancelation messages is critical since a delayed feedback cancelation is very likely to provoke a feedback implosion.

3.2   Message Piggybacking

The feedback requests and the cancelation messages from the sender can both be piggybacked on data packets to minimize network overhead. In case a unicast feedback channel is used, piggybacking has to be done with great care since at low sending rates the delayed cancelation messages may provoke a feedback implosion. This undesired behavior is likely to occur when the inter-packet spacing between data packets gets close to the maximum feedback delay.

The problem can be prevented by not piggybacking but sending a separate cancelation message at low data rates (i.e., introducing an upper bound on the amount of time by which a cancelation message can be delayed). If separate cancelation messages are undesirable, it is necessary to increasing the maximum feedback delay T in proportion to the time interval between data packets.

3.3   Removing Latency Bias

Plain exponential feedback favors low-latency receivers since they get the feedback request earlier and are thus more likely to suppress other feedback. In case the receivers know their own latency τ as well as an upper bound on the latency for all receivers τ_{max}, it is possible to remove this bias. Receivers simply schedule the sending of the feedback message for time $t + (\tau_{max} - \tau)$ instead of t.

In fact, this unbiasing itself introduces a slight bias *against* low-latency receivers in case unicast feedback channels are used. While the first feedback message is unaffected, subsequent duplicates are more likely to come from high-latency receivers, since they will receive the feedback suppression notification from the sender later in time.

If it is not necessary to remove the latency bias, the additional receiver heterogeneity generally improves the suppression characteristics of the feedback mechanism, as demonstrated in [9]. Similar considerations hold for the suppression mechanisms discussed in the following sections.

4   Extremum Detection

Let us now consider the case where not only an arbitrary response from the group is required but an extreme value for some parameter from within a group. Depending on the purpose the required extremum can be either a maximum or a minimum. Without loss of generality we will formulate all algorithms as maximum detection algorithms.

4.1   Basic Extremum Detection

An obvious approach to introduce a feedback preference scheme is to extend the normal exponential feedback mechanism with the following algorithm:

Algorithm 2. *(Basic Extremum Detection):*
Let $v_1 > v_2 > \ldots > v_k > 0$ be the set of response values of the receivers.

Upon receipt of a feedback request each receiver sets a feedback timer according to Algorithm 1. When a receiver with value v is notified that another receiver already gave feedback with $v' \geq v$, it cancels its timer. Otherwise, when the feedback timer expires (i.e., for all previous notifications $v' < v$ or no notifications were received at all), the receiver sends a feedback message with value v.

With this mechanism the sender will always obtain feedback from the receiver with the largest response value within one feedback round.

Let us now analyze the algorithm in detail: Following Equation (3) we use n for the actual number of potential responders and denote the expected number of feedback messages in Algorithm 1 with $R(n) := E[M]$. Let p_i be the fraction of responders with value v_i. For $k = 1$ the problem reduces to Algorithm 1 and we expect $R(n)$ feedback messages. For $k = 2$ we can reduce the problem to the previous case by assuming that every v_1 responder responds with both a v_1 and a v_2

message. Hereby, we can treat both groups independently from each other while preserving the fact that v_1 responders also stop further (unnecessary) responses from v_2 responders. Summing up both expected values we have $R(p_1 n) + R(n)$ messages. However, p_1 of the v_2 messages were sent by v_1 responders and are thus duplicates. Subtracting these duplicates we obtain $R(p_1 n) + p_2 R(n)$ for the expected number of responses.

This argument can be extended to the general case

$$E[M] = R(p_1 n) + \frac{p_2}{p_1 + p_2} R(p_1 n + p_2 n)$$

$$+ \frac{p_3}{p_1 + p_2 + p_3} R(p_1 n + p_2 n + p_3 n)$$

$$+ \cdots + p_k R(n)$$

$$= \sum_{i=1}^{k} \frac{p_i}{P_i} R(P_i n) \tag{4}$$

where $P_i := p_1 + p_2 + \ldots + p_i$ and thus $P_k = 1$. According to [6], $R(n)$ remains approximately constant over wide ranges of n. Assuming $R(n) \simeq R$, $p_i \simeq \frac{1}{k}$, and $k \gg 1$ we have

$$E[M] \simeq \left(1 + \frac{1}{2} + \frac{1}{3} + \cdots + \frac{1}{k}\right) R$$

$$\simeq (\ln k + C)R \tag{5}$$

where $C = 0.577\ldots$ denotes the Euler constant.

From this analysis we see that the number of possible feedback values has an impact on the expected number of feedback messages. For a responder set with a real-valued feedback parameter this results in $E[M] \simeq \ln(n)R$.

4.2   Class-Based Extremum Detection

Although this logarithmic increase is well acceptable for a number of applications, the algorithm's properties can be further improved by the introduction of feedback classes. Within those classes no differentiation is made between different feedback values. It is not necessary to choose a fixed size for all classes. The class size can be adapted to the required granularity for certain value ranges. In case a fixed number of classes is used, the expected number of feedback messages increases only by a constant factor over normal exponential feedback. This increase is expectedly observed in the simulation results shown in Figure 1. As the number of classes approaches the number of receivers, the increase in feedback messages follows more and more the logarithmic increase for real-valued feedback as stated in Equation 5. For all simulations in this paper we use the parameters $N = 100,000$ and $T = 4\tau$ and average the results over 200 simulation runs, unless stated otherwise.

By adjusting the classes' positions depending on the actual value distribution, the number of classes required to cover the range of possible feedback values can

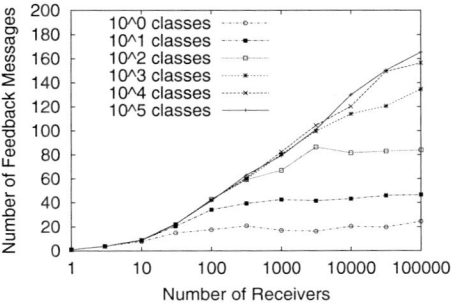

Fig. 1. Uniformly Sized Classes.

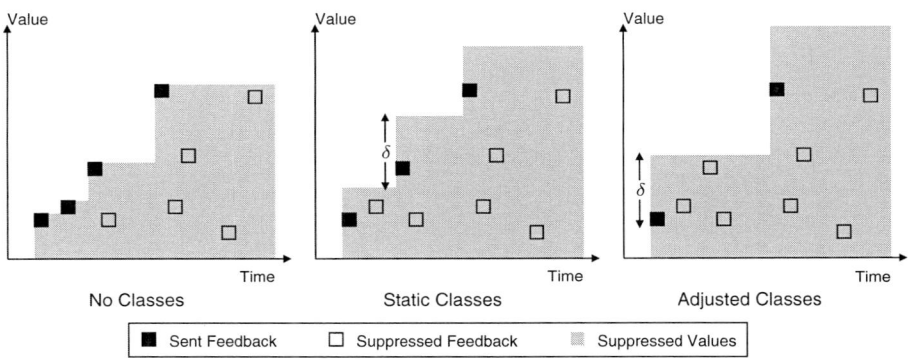

Fig. 2. Class-Based Suppression with Variable Class Position.

be reduced without increasing the intervals' actual size. Thereby, the granularity of the feedback suppression (i.e., to what extent less optimal values can suppress better values) remains unchanged while the number of feedback messages is reduced.

Figure 2 gives a schematic overview of this mechanism. The first diagram shows the classless version of the feedback algorithm. Here, each time a feedback message v_i is sent, the range of suppressed value increases to $[0; v_i]$. A total of four feedback messages is sent in this example. The second diagram shows the same distribution of feedback for the case of static classes. We assume equally sized classes of size δ and $v_1 \in [0; \delta]$ for this example. After receipt of the first feedback message v_1, the entire range $[0; \delta]$ of the lowest feedback class is suppressed. Only when a value outside this class is to be reported another message is sent, resulting in three feedback messages in total. The third diagram shows the case of dynamically adjusted classes. Upon receipt of the first feedback message v_1 the suppression limit is immediately raised to $v_1 + \delta$ and thus the value range $[0; v_1 + \delta]$ is now being suppressed. Through this mechanism feedback is reduced to only two messages.

With the above considerations, an elegant way to introduce feedback classes is the modification of Algorithm 2 to suppress feedback not only upon receipt of values strictly larger than the own value v but also upon receiving values $v' \geq (1 - q)v$. This results in an adaptive feedback granularity dependent on the absolute value of the optimum.

Algorithm 3. *(Adaptive Class-Based Extremum Detection):*
Let q be a tolerance factor with $q \in [0; 1]$. Modify Algorithm 2 such that a responder with value v cancels its timer if another responder has already sent feedback for value v' with $v' \geq (1 - q)v$.

For $q = 0$ the algorithm is equivalent to Algorithm 2, whereas for $q = 1$ we obtain Algorithm 1.

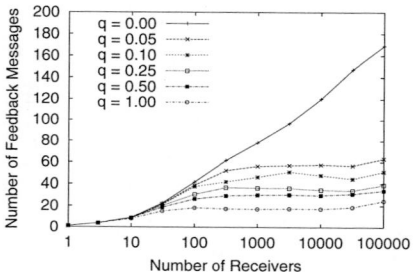 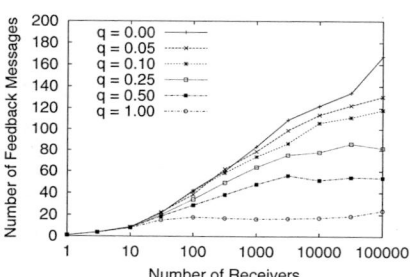

Fig. 3. Number of Feedback Messages for Maximum Search (left) and Minimum Search (right).

Assuming the values v_i to be evenly distributed between rv_{max} and v_{max} $(0 < r < 1)$ we have approximately k feedback classes[3], where $k < \frac{\ln r}{\ln 1 - q}$. For a value range $0 < v_i < 1$ we can assume $k < \frac{-\ln n}{\ln 1 - q}$, setting r inversely proportional to the number of receivers since the receiver set is too small to cover the whole range of possible values.

Approximating further with

$$p_i = \frac{(1-q)^{i-1} - (1-q)^i}{1-r} = q\frac{(1-q)^{i-1}}{1-r}$$

and

$$P_i = \frac{1 - (1-q)^i}{1-r}$$

---

[3] We assume the parameter range $(r, 1)$ to be fully covered by the feedback classes which is not strictly the case for this algorithm. This approximation thus overestimates the expected number of feedback messages.

we have

$$E_{max}[M] < qR \sum_{i=1}^{k} \frac{(1-q)^{i-1}}{1-(1-q)^i} \tag{6}$$

The mechanism strongly benefits from the feedback classes being wider near the maximum and so holding more values than the classes near the minimum. As a consequence, the expected number of feedback messages is much lower compared to that of the previous algorithm. Note that for small r, the number of members with $v < (1-q)^{-1}r$ can be very small. Eventually, these feedback classes will contain only a single member and we therefore loose the desired suppression effect that leads to a sub-logarithmic increase of feedback messages. In maximum search this effect cannot be observed since already a single response in the larger feedback classes near the maximum will suppress all feedback from the potentially large number of small classes. In fact, this characteristic is not specific to maximum and minimum search but rather depends on the classes being large or small near the optimum.

To demonstrate the effect we will calculate the expected number of feedback messages for a minimum search scenario: The feedback values v_i are again evenly distributed between rv_{max} and v_{max}, but in contrast to Algorithm 3 a responder cancels its timer if a response with $v' \le (1-q)v$ is received. The algorithm produces the minimal value of the group within a factor of q.[4]

The feedback classes are in the opposite order as compared to our previous calculation.

$$p_i = \frac{(1-q)^{k-i} - (1-q)^{k-i+1}}{1-r} = q \frac{(1-q)^{k-i}}{1-r}$$

and

$$P_i = \frac{(1-q)^{k-i} - (1-q)^k}{1-r}$$

Thus

$$E_{min}[M] < qR \sum_{i=1}^{k} \frac{1}{1-(1-q)^i} \tag{7}$$

$$\approx (1-q)\, E_{max}[M] + kqR$$

Hence, for small r (large k) the sum is significantly larger than in the previous case.

Both scenarios have been simulated with various values for q. The sub-logarithmic increase of feedback messages can be seen in both plots shown in Figure 3. But only with maximum search where the feedback-classes near the search goal are wider, the strong class-induced suppression dominates the $\ln(n)$ scale-effect.

---

[4] As far as the expected number of feedback messages is concerned, this mechanism is equivalent to a maximum search with small class sizes for classes close to the optimum.

66   Jörg Widmer and Thomas Fuhrmann

Numeric values for the upper limits on the expected number of feedback
messages in both scenarios can be obtained from Equations (6) and (7). Some
example values are shown in Table 1. These limits match well with the results
of our simulations.

Table 1. Upper Limits (as Factor of R) for the Expected Number of Feedback
Messages ($r = 10^{-2}$).

q	0.05	0.10	0.25	0.50	1.00
Maximum search	3.64	3.00	2.19	1.59	1.00
Minimum search	7.91	7.00	5.64	3.80	1.00

As mentioned before, Algorithm 3 guarantees a maximum deviation from the
true optimum of a factor of q. It is worthwhile to note that this factor really
is an upper bound on the deviation. Almost always the reported values will
be much closer to optimal since the sender can choose the best one of all the
responses given. The deviation of the best reported value from the optimum for
different tolerance factors q is depicted in Figure 4. On average, with normal
exponential suppression (i.e., $q = 100\%$) the best reported value lies within 10%
of the optimum, for $q = 50\%$ the deviation drops to less than 0.15%, for $q = 10\%$
we obtain less than 0.02% deviation, etc. Thus, even for relatively high q with
consequently only a moderate increase in the number of feedback messages, the
best feedback values have only a marginal deviation from the optimum.

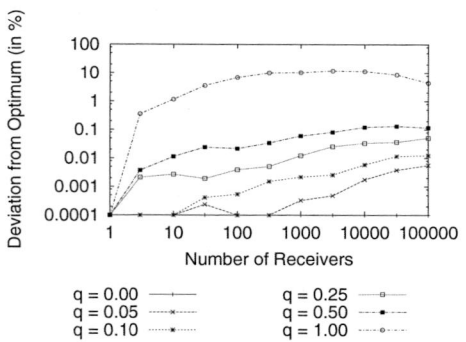

Fig. 4. Feedback Quality with Different Tolerance Values.

5   Biased Feedback

The previously described algorithms yield considerable results for various cases of extremum detection. However, they will not affect the expected value of the first feedback message but only improve the expected values of subsequent messages. In certain cases, the algorithms can be further improved by biasing the feedback timers. Increasing the probability that $t_1 < t_2$ if $v_1 > v_2$ results in better feedback behaviour but we must carefully avoid a feedback implosion for cases where many large values are present in the responder group. Without loss of generality we assume $v \in [0, 1]$ for the remainder of this section.

If the probability distribution of the values v is known, the number of responses can be minimized using the following algorithm:

Algorithm 4. *(Deterministic Feedback Suppression):*
Let $P(v) = P(v' < v)$ be the probability distribution function of the values v within the group of responders. We follow Algorithm 1, but instead of drawing a random number we set the feedback time directly to

$$t = T \max(0; 1 + \log_N(1 - P(v)))$$

Clearly, duplicate feedback responses are now only due to network latency effects since the responder with the maximum feedback value is guaranteed to have the earliest response time. However, the feedback latency is strongly coupled to the actual set of feedback values. Moreover, if the probability distribution of this specific set does not match the distribution used in the algorithm's calculation, feedback implosion is inevitable. For this reason, Algorithm 4 should only be used if the distribution of feedback values is well known for each individual set of values.

The latter condition is crucial. In general, it does not hold for values from causally connected responders. Consider for example the loss rate for multicast receivers: If congestion occurs near the sending site, all receivers will experience a high packet loss rate simultaneously. Since the time-average distribution does not show this coherence effect the algorithm presented above will produce feedback implosion, if used to solicit responses from high-loss receivers. Due to this effect, the application of this simple mechanism is quite limited. It can be used, for example, with application level values where no coherence is generated within the network.

A simple way to adopt the key idea of value-based feedback bias is to mix value-based response times with a random component. This mechanism can be applied in various cases where coherence effects prohibit the application of Algorithm 4. Let us study an example:

Algorithm 5. *(Feedback with Combined Bias):*
Apply Algorithm 1 but modify the feedback time to

$$t = T \max(0; (1 - v) + v(1 + \log_N x))$$
$$= T \max(0; (1 + \log_N x^v)) \tag{8}$$

Here, the feedback time consists of a component linearly dependent on the feedback value and a component for the exponential feedback suppression. The feedback time t is increased in proportion to decreasing feedback values v and a smaller fraction of T is used for the actual suppression. As long as at least one responder has a sufficiently early feedback time to suppress the majority of other feedback this distribution of timer values greatly decreases the number of duplicate responses while at the same time increasing the quality of the feedback (i.e., the best reported value with respect to the actual optimum value of the receiver set). Furthermore, in contrast to pure extremum detection algorithms this mechanism improves the expected feedback value of the first response as well as subsequent responses.

However, the feedback suppression characteristics of the above mechanism still depend at least to some extent on the value distribution at the receivers. Some extreme cases such as $v = 0$ for all receivers will always result in a feedback implosion. A more conservative approach is to not combine bias and suppression but use a purely additive bias.

Algorithm 6. *(Feedback with Additive Bias):*
Apply Algorithm 1 but modify the feedback time to

$$t = T \max\left(0; \gamma(1 - v) + (1 - \gamma)(1 + \log_N x)\right) \qquad (9)$$

with $\gamma \in [0; 1]$.

To retain the same upper bound on the maximum feedback delay, it is necessary to split up T and use a fraction of T to spread out the feedback with respect to the response values and the other fraction for the exponential timer component. As long as $(1 - \gamma)T$ is sufficiently large compared to the network latency τ, an implosion as in the above example is no longer possible.

To better demonstrate the characteristics of these modifications, Figures 5 to 7 show how the feedback time changes with respect to response values compared to normal unbiased feedback according to Algorithm 1. A single set of random variables was used for all the simulations to allow a direct comparison of the results. For the simulations, the parameters N and n were set to $10,000$ and $2,000$ respectively.[5] In these simulations we do *not* consider maximum search but only how feedback biasing affects the distribution of feedback timers. Thus, to isolate the effect of feedback biasing, only a single feedback class was used such that the first cancelation notification suppresses *all* subsequent feedback. All simulations were carried out with $T = 4\tau$ as well as $T = 8\tau$ to demonstrate the impact of the feedback delay on the number of feedback responses. Each graph shows the feedback times in τ of the receiver set along the x-axis and the corresponding response values on the y-axis for each of the three feedback mechanisms *no bias* (Algorithm 1), *combined bias* (Algorithm 5), and *additive*

---

[5] Note that using $n = N = 10,000$ instead of $n = 2,000$ would *reduce* the probability of an implosion since the probability that one early responder suppresses all others increases.

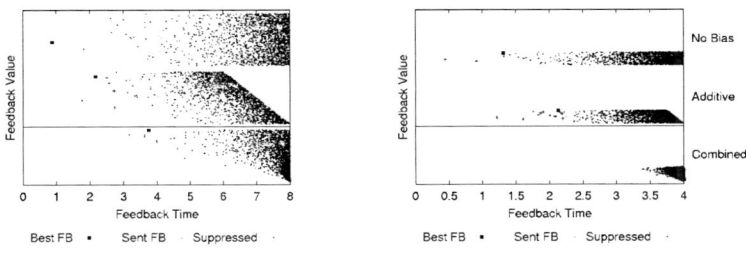

Fig. 5. Feedback Time and Value (Uniform Distribution of Values).

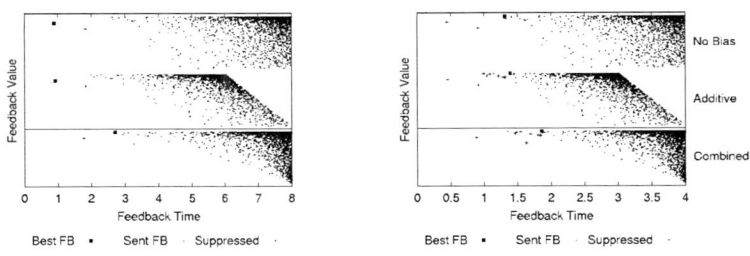

Fig. 6. Feedback Time and Value (Exponential Distribution of Values).

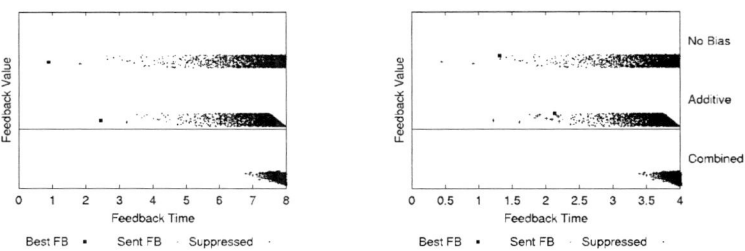

Fig. 7. Feedback Time and Value (Truncated Uniform Distribution of Values).

bias (Algorithm 6) with $\gamma = 1/4$. Suppressed feedback messages are marked with a dot, feedback that is sent is marked with a cross, and the black square indicates which of these feedback messages had a value closest to the actual optimum of the receiver set.

In the graphs in Figure 5, the response values of the receivers are uniformly distributed. When no feedback bias is used, the first response that suppresses the other responses is random in value. In contrast, both feedback biasing methods result in the best reported feedback value being very close to the actual optimum. The number of sent feedback messages is higher with the two biasing methods since a smaller fraction of T is used for feedback suppression. Naturally, the number of feedback messages also increases when T is smaller as depicted in the right graph.

In Figure 6, the same simulations were carried out for an exponential distribution of response values with a high probability of being close to the optimum. (When a reversed exponential distribution with most values far from the optimum is used, the few good values suppress all other feedback and again a feedback implosion is always prevented.) As can be seen from the graph, feedback suppression works well even when the actual distribution of response values is no longer uniform. For a uniform as well as an exponential distribution of response values, the *combined bias* suppression method results in fewer feedback messages while maintaining the same feedback quality.

However, as mentioned before, combining bias and suppression permits a feedback implosion when the range of feedback values is smaller than anticipated. In this case, the bias results in an unnecessary delaying of feedback messages, thus reducing the time that can be used for feedback suppression. In Figure 7, the response values are distributed uniformly in $[0; 0.25]$ instead of $[0; 1]$. For $T = 4\tau$, the time left for feedback suppression is τ, resulting in a scenario where no suppression is possible and each receiver will send feedback. Even when $T = 8\tau$ and thus a time of 2τ can be used for the feedback suppression, the number of feedback messages is considerably larger than in simulations with an additive bias. The exact numbers for the feedback responses of the three methods are given in Table 2.

Table 2. Number of Responses with the Different Biasing Methods.

Feedback Time	No Bias	Additive	Combined
T=4, Uniform	5	19	15
T=4, Exponential	5	14	12
T=4, Truncated	5	14	2000
T=8, Uniform	2	6	4
T=8, Exponential	2	2	2
T=8, Truncated	2	2	334

For suppression to be effective, the amount of time reserved for the exponential distribution of the feedback timers should not be smaller than 2τ. Thus, the feedback implosion with Algorithm 5 can be prevented by bounding v such that $vT > 2\tau$ (i.e., using $v' = \max(v; 2\tau/T)$ instead of v in Equation 8). In the worst case, the distribution of the feedback timers is then similar to an unbiased exponential distribution with $T = 2\tau$. A higher upper bound can be used to reduce the expected number of feedback messages in the worst case. The same considerations hold for the choice of the value of γ for the additive feedback bias.

The outcome of a single experiment is not very representative since the number of feedback messages is extremely dependent on the feedback values of the early responders. As for the previously discussed feedback mechanisms, we depict the number of feedback messages for combined and additive bias averaged over 200 simulations in Figure 8.

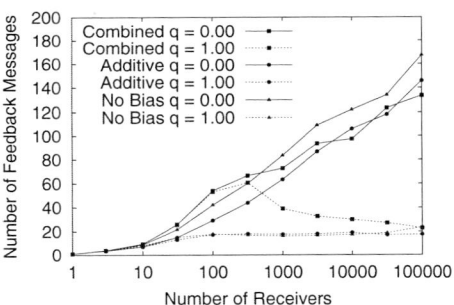

Fig. 8. Number of Responses with Feedback Biasing.

The main advantage of the feedback bias is that the expected response value for early responses is improved. This not only reduces the time until close to optimal feedback is received (with unbiased feedback and class-based suppression, close to optimal feedback is likely to arrive at the end of a feedback round) but also reduces the number of responses with less optimal feedback.

Figure 9 shows how the feedback quality improves compared to the normal exponential feedback suppression when biasing the feedback timer. The maximum deviation is reduced from about 15% to 6% for additive bias and to less than 2% for combined bias.

While a similar increase in feedback quality can be achieved by using feedback classes (at the expense of an increased number of feedback messages) only with a feedback bias is it possible to improve the quality of the *first* feedback message. In case a close to optimal value is needed very quickly, using either Algorithm 5 or Algorithm 6 can be beneficial. Figure 10 depicts the average deviation of the value of the first feedback message from the optimum. Here, the increase in quality is much more obvious than in the previous case. With all unbiased

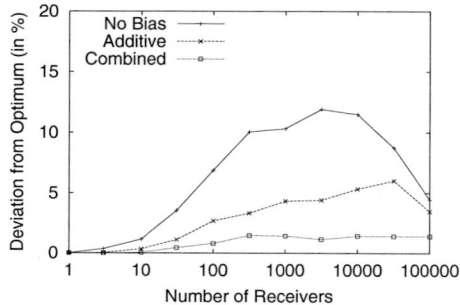

Fig. 9. Deviation of Best Responses Value from Optimum.

feedback mechanisms, the first reported value is random and thus the average deviation is 50% (for large enough n) whereas the combined and the additive biased feedback mechanisms achieve average deviation values around 10% and 30% respectively.

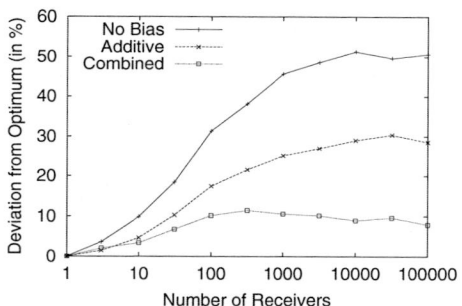

Fig. 10. Deviation of First Response Value from Optimum.

Lastly, the expected delay until the first feedback message is received is of concern. While all mechanisms adhere to the upper bound of T, feedback can be expected earlier in most cases. In Figure 11 we show the average feedback delay for biased and unbiased feedback mechanisms. For all algorithms the feedback delay decreases logarithmically for an increasing number of receivers. The exact run of the feedback curve depends on the amount of time used for suppression. For this reason, unbiased feedback delay drops faster than biased feedback, since a bias can only delay feedback messages compared to unbias feedback. In case the number of receivers is estimated correctly (i.e., $n = N$), the feedback delay for unbiased feedback drops to τ, the minimum delay possible for such a feedback system. Biased feedback delay is slightly higher with approximately 1.5τ.

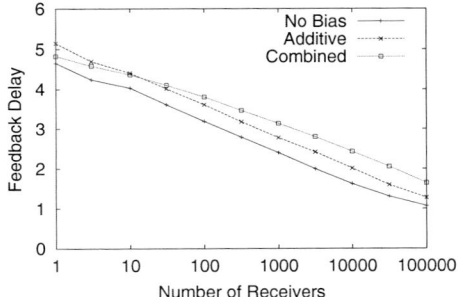

Fig. 11. Average Feedback Delay.

6   Conclusions

In this paper we presented mechanisms that improve upon the well-known concept of exponential feedback suppression in case feedback of some extreme value of the group is needed. We discuss two orthogonal methods to improve the quality of the feedback given. If no information is available about the distribution of the values at the receivers, a safe method to obtain better feedback is to modify the suppression mechanism to allow the sending of high valued feedback even after a receiver is notified that some feedback was already given. We give exact bounds for the expected increase in feedback messages for a given improvement in feedback quality. If more information about the distribution of feedback values is available or certain worst-case distributions are very unlikely, it is furthermore possible to bias the feedback timer. The better the feedback value the earlier the feedback is sent, thus suppressing later feedback with less optimal values. The modified suppression mechanism and the feedback biasing can be used in combination to further improve the feedback process.

The mechanisms discussed in this paper have been included in the TCP-friendly Multicast Congestion Control Protocol (TFMCC) [14]. It uses class-based feedback cancelation as well as feedback biasing to determine the current limiting receiver (i.e., the receiver with the lowest expected throughput of the multicast group). The protocol depends on short feedback delays in order to quickly respond to congestion. Selecting the correct receiver as current limiting receiver is critical for the functioning of the protocol since a wrong choice may compromise the TCP-friendly characteristics of TFMCC. In that sense, the feedback mechanism is an important part of the TFMCC protocol.

Extremum feedback is not yet included in any other application, but we believe a number of applications can benefit from it.

6.1   Future Work

In the future we would like to continue this work in several directions.

Most applications need to consider only one type of feedback value. Nevertheless, it may sometimes be useful to get multivalued feedback, for example to monitor some critical parameters of a large network, where changes in each of the parameters are equally important. It may not always be possible to aggregate different types of values to one single "ranking" value. In this case, a multivalued feedback mechanism clearly has better suppression characteristics than separate feedback mechanisms for each of the relevant values.

Another important step will be the combination of knowledge about the value distribution within the responder group with implosion avoidance features. Several mechanisms to estimate the size of the receiver set from the feedback time and the number of feedback messages with exponential feedback timers have been proposed [8,5]. Combining such estimation methods with extremum feedback, it should be possible to estimate the distribution of response values at the receivers in case this distribution is not known. For continuous feedback, this knowledge can then be used to generate feedback mechanisms based on Algorithm 4. In scenarios where the distribution of response values is not uniform, we expect that such an approach will outperform the biasing mechanisms presented in section 5, which do not take the distribution into account.

Taking these considerations one step further, in some cases the maximum change of the relevant state during one feedback round is bounded. For example, in the case of TFMCC, the measurements to determine round-trip time and loss event rate are subject to smoothing, thus limiting the maximum rate increase and decrease per round-trip time. In case some information about the previous distribution of feedback values is available (e.g., from the previous feedback round), it is possible to infer the worst case distribution of the current feedback round. This allows to further improve the feedback algorithm by tailoring it to the specific distribution.

Acknowledgments

We would like to thank Mark Handley who inspired the work on feedback suppression and we thank the anonymous reviewers for their helpful comments. Furthermore, Thomas Fuhrmann would like to thank Ernst Biersack for supporting this work by hosting him during a sabbatical stay at Eurecom.

References

1. BASU, A., AND GOLESTANI, J. Architectural issues for multicast congestion control. In *International Workshop on Network and Operating System Support for Digital Audio and Video (NOSSDAV)* (June 1999).
2. BOLOT, J.-C., TURLETTI, T., AND WAKEMAN, I. Scalable feedback control for multicast video distribution in the Internet. *Proceedings of the ACM SIGCOMM* (Sept. 1994), 58 – 67.
3. DONAHOO, M. J., AND AINAPURE, S. R. Scalable multicast representative member selection. In *IEEE INFOCOM* (March 2001).
4. FENNER, B., HANDLEY, M., HOLBROOK, H., AND KOUVELAS, I. Protocol independent multicast - sparse mode (pim-sm): Protocol specification, Nov. 2000. Internet draft draft-ietf-pim-sm-v2-new-01.txt, work in progress.
5. FRIEDMAN, T., AND TOWSLEY, D. Multicast session membership size estimation. In *IEEE Infocom* (New York, NY, Mar. 1999).
6. FUHRMANN, T., AND WIDMER, J. On the scaling of feedback algorithms for very large multicast groups. *Computer Communications 24*, 5-6 (Mar. 2001), 539 – 547.
7. JUANG, J. Y., AND WAH, B. W. A unified minimum-search method for resolving contentions in multiaccess networks with ternary feedback. *Information Sciences 48*, 3 (1989), 253–287. Elsevier Science Pub. Co., Inc., New York, NY.
8. LIU, C., AND NONNENMACHER, J. Broadcast audience estimation. In *IEEE Infocom* (Tel Aviv, Israel, Mar. 2000), pp. 952–960.
9. NONNENMACHER, J., AND BIERSACK, E. W. Scalable feedback for large groups. *IEEE/ACM Transactions on Networking 7*, 3 (June 1999), 375 – 386.
10. RIZZO, L. pgmcc: A TCP-friendly single-rate multicast congestion control scheme. In *Proc. ACM SIGCOMM* (Stockholm, Sweden, August 2000), pp. 17 – 28.
11. SCHULZRINNE, H., CASNER, S., FREDERICK, R., AND JACOBSON, V. Rtp: A transport protocol for real-time applications. *RFC 1889* (January 1996).
12. WAH, B. W., AND JUANG, J. Y. Resource scheduling for local computer systems with a multiaccess network. *IEEE Trans. on Computers C-34*, 12 (Dec. 1985), 1144–1157.
13. WIDMER, J., AND FUHRMANN, T. Extremum feedback for very large multicast groups. Tech. Rep. TR 12-2001, Praktische Informatik IV, University of Mannheim, Germany, May 2001.
14. WIDMER, J., AND HANDLEY, M. Extending equation-based congestion control to multicast applications. In *Proc. ACM SIGCOMM* (San Diego, CA, Aug. 2001).

An Overlay Tree Building Control Protocol*

Laurent Mathy[1], Roberto Canonico[2], and David Hutchison[1]

[1] Lancaster University, UK
{laurent,dh}@comp.lancs.ac.uk
[2] University Federico II, Napoli, Italy
roberto.canonico@unina.it

Abstract. TBCP is a generic Tree Building Control Protocol designed
to build overlay spanning trees among participants of a multicast session,
without any specific help from the network routers. TBCP therefore falls
into the general category of protocols and mechanisms often referred to
as Application-Level Multicasting. TBCP is an efficient, distributed protocol that operates with partial knowledge of the group membership
and restricted network topology information. One of the major strategies in TBCP is to reduce convergence time by building as good a tree
as possible early on, given the restricted membership/topology information available at the different nodes of the tree. We analyse our TBCP
protocol by means of simulations, which shows its suitability for purpose.

1 Introduction

It is well known that the research community has proposed several basic models
and a plethora of supporting protocols for multicast in the Internet, none of
which has been ultimately deployed on the very large scale of the whole Internet [3]. Making multicast deployment a more difficult task is the fact that even
the network layer service of the Internet is in the process of being (at least partially) updated, due to the introduction of a new version of the Internet Protocol
(IPv6).

Both the versions of the IP protocol support multicast transmission of data-
grams to several receivers [2], i.e. they reserve a set of addresses to identify
groups of receivers and they assume that network routers are able to replicate
packets and deliver them to all entities registered as members of a group. To
carry out such a job, a number of complementary control and routing protocols
must be deployed both in the end systems and in the network routers. Such
deployment of new software in the network routers (i.e. changes to the network
infrastructure) is a major hurdle to ubiquitous deployment, and is responsible
for very long rollout delays.

This situation has led the research community to propose mechanisms and
protocols to build overlay spanning trees among hosts in a multicast session
(and possibly service nodes, which are hosts placed strategically in the network

---

* This work was supported by the EC IST GCAP (Global Communication Architecture and Protocols) Project.

J. Crowcroft and M. Hofmann (Eds.): NGC 2001, LNCS 2233, pp. 76–87, 2001.

to facilitate and optimise the construction of these overlays). Obviously, the distribution of multicast data with such overlay spanning trees is less efficient than in native IP multicast mode, but this relative performance penalty must be contrasted with the easy and speed of deployment offered by the overlay technique.

Until recently, the main paradigm enabling multicast in the Internet was the dual concept of group and associated group address. However, these concepts alone lead to an open service model for IP multicast where anybody can send to a group and that therefore presents serious scalability, security and deployment concerns [3]. This is why a restricted Internet multicast model called Source-Specific Multicast (SSM) has been proposed [8][5]. SSM is based on the concept of multicast channel (i.e. a pair (source address, multicast address)) and the fact that only the channel source can send on the channel. This new model for multicast service, because of its simplicity and elegance, has attracted widespread support within both the academic and industrial communities.

Unfortunately, this SSM model breaks many of the mechanisms and protocols that have been proposed for reliable multicast communications (e.g. [6]), due to the lack of a multicast backchannel from the receivers to the rest of the multicast group. It should be noted that such a problem is not specific to an SSM multicast model, as it can also manifest itself on asymmetric satellite link with terrestrial returns, some firewall configurations and some proprietary multicast deployment schemes (e.g. UUnet multicast deployment).

This observation, along with the fact that Tree-based ACKs (TRACKS) [16] [15][14] appear to be an excellent candidate for scalable, (near) real-time reliable multicast communications, argues for overlay multicast spanning trees to be used for control purposes in a reliable multicast scenario.

In this paper, after briefly reviewing techniques that have been proposed to build both multicast data and control trees, we present our Tree Building Control Protocol (TBCP) which is capable of efficiently building both data and control trees in any network environment. Finally, we analyse the performance and characteristics of the overlay trees built by TBCP.

2   Related Work

The concept of tree has proved to be a particularly well suited and efficient dissemination structure for carrying both data and control information in group communications.

Even if, and when, IP multicast becomes widely deployed, the need for tree-based control structures will remain in a reliable multicast scenario. The nodes of such control trees must understand and take part in the control protocols used. These nodes can either be the end-hosts (sender and receivers), service nodes, network routers implementing the appropriate control functions, or any combination of these.

The best way to build control trees that are congruent to the IP multicast data distribution tree is, of course, to support the construction of the control tree within the network. Pretty Good Multicast (PGM) [4] builds control trees based on the concept of *reverse path*. PGM control trees are not overlay trees because internal tree nodes are PGM routers. For that reason, PGM is very efficient.

In [11], overlay trees are built based on *positional reachcasting*. Reachcasting consists of sending packets to the nearest member host in any direction of a multicast routing tree and relies on multicast routers supporting the notion of nearest-host routing. The overlay trees built by with this technique are optimal.

Unfortunately, the above mentioned tree building techniques require modifications to the network infrastructure and are yet to be widely deployed.

Techniques to build overlay trees without relying on special router features, apart from multicast routing, have also been proposed. TMTP [16] uses *expanding ring searches (ERS)* to discover tree neighbours. However, ERS does not perform well in asymmetrical networks, so that the notion of *unsolicited neighbour announcements* has been introduced in TRAM [10]. TMTP and TRAM both require multicast support in the network and will not operate in an SSM context.

More recently, techniques for building overlay spanning trees without depending on multicast routing support in the network have been proposed. We do not attempt an exhaustive review of this area, but rather try and present the most significant proposals.

ALMI [13] is representative of centralised techniques, where a *session controller* has total knowledge of group membership as well as of a the characteristics of a mesh connecting the members (this mesh improves over time). Based on this knowledge, the session controller builds a spanning tree whose topology is distributed to all the members. However, as centralised approach do not scale, distributed techniques are required to support groups larger than a few tens of members.

In Yoid [7], a prospective receiver learns about some tree nodes from a rendez-vous point and chooses one of them as its parent. The place in the tree where a new member joins may not be optimal and a distributed tree management protocol is employed to improve the tree over time. However, for scalability reasons, the convergence time to optimality can be rather slow. Yoid hosts also maintain an independent mesh for robustness.

In Narada [1], the hosts first build a random mesh between themselves, then a (reverse) shortest path spanning the mesh. For robustness, each host gains full knowledge of the group membership (Narada is targetted towards small multicast groups) through gossiping among mesh neighbours, and this knowledge is used to slowly improve the quality of the mesh.

Overcast [9] is another approach aimed at an infrastructure overlay. It is specifically targeted towards bandwidth efficient overlay trees. Overcast is a tree-first approach building *unconstrained* trees (i.e. tree nodes do not limit the number of children they serve).

3   Tree Building Control Protocol (TBCP)

Building an overlay spanning tree among hosts presents a significant challenge. To be deployable in practice, the process of building the tree should not rely on any network support that is not already ubiquitous. Furthermore, to ensure fast and easy deployment, we believe the method should avoid having to rely on "its own infrastructure" (e.g. servers, "application-level" routing protocols, etc.), since the acceptance and deployment of such infrastructure could hamper the deployment of the protocol relying on it. Consequently, one of our major requirements is to design a tree building control protocol that operate between end-systems exclusively[1], and considers the network as a "black-box". End-systems can only gain "knowledge" about the network through host-to-host measurement samples.

From a scalability point of view, it is also unrealistic to design a method where pre-requisite knowledge of the full group membership is needed before the spanning tree can be computed. This is especially true in the context of multicast sessions with dynamic group membership. For robustness and efficiency, as well as scalability, it is also preferable to avoid making use of a centralised algorithm.

Our goal is therefore to design a distributed method which builds a spanning tree (which we also call the *TBCP tree*) among the hosts of a multicast session without requiring any particular interactions with network routers nor any knowledge of the network topology, and which does so with only partial knowledge of the group membership.

TBCP is a tree-first, distributed overlay spanning tree building protocol, whose strategy is to place members in a (near) optimal position at joining time.

Compared with the proposals discussed in the previous section, TBCP can be viewed as complementing Yoid, in the sense that TBCP could advantageously replace the rather "random bootstrap" approach in Yoid. Unlike Narada, TBCP does not require full membership knowledge and is not a mesh-first protocol. Also, because of the fundamental strategy in TBCP, tree convergence time should be much faster than in Yoid and Narada.

The general approach to building the tree in TBCP is similar to the one followed in overcast. However, TBCP is not restricted to bandwidth optimised trees and, maybe more importantly, TBCP trees are explicitly constrained (each host fixes an upper-limit on the number of children it is willing to support).

3.1   TBCP Join Procedure

In TBCP, the root of the spanning tree (e.g. main sender to the group) is used as a rendez-vous point for the associated TBCP tree, that is new nodes "join" the tree at the root. Hence a TBCP tree can be identified by the (S,SP) pair, where S is the IP address of the TBCP tree root, and SP the port number used by

---

[1] We want to emphasise that infrastructure nodes are not considered as a pre-requisite in our approach. However, should such nodes exist, TBCP *could*, of course, be used to interconnect them.

the root for TBCP signalling operations. This information is the only advertised information needed to join the TBCP tree.

Each TBCP entity (including the tree root) fixes the maximum number of "children" it is willing to accommodate in the spanning tree. This value, called the *fanout* of the entity, allows the entity to control the load of traffic flowing on the TBCP tree that it will handle. The TBCP Join procedure is a "recursive"

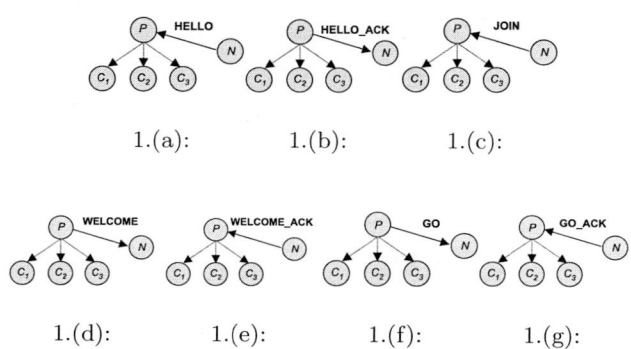

1.(a):   1.(b):   1.(c):

1.(d):   1.(e):   1.(f):   1.(g):

Fig. 1. TBCP Join Procedure Messages.

mechanism and works as follows:

1. A newcomer N contacts a candidate parent P with a HELLO message, starting with the tree root S (figure 1.(a)).
2. P sends back to N the list of its existing children C_i in a HELLO_ACK message (figure 1.(b)), starts a timer T_0 and wait for a JOIN message from N (figure 1.(a). This timer is needed because, for consistency reasons, P cannot accept any new HELLO message until it has finished dealing with the current newcomer.
3. N estimates its "distance" (i.e. takes measurement samples) from P and all C_is and sends this information to P in a JOIN message (figure 1.(c)). Note that if P has not received this JOIN message within its timer T_0, P sends a RESET message to N, meaning that N needs to restart the procedure from stage 1.
4. P finds a place for N, by evaluating all possible "local" configurations having P as parent and involving $C_i \cup N$ (see figure 2). A score function is used to evaluate "how good" each configuration is, based on the distance estimates among P, N and the C_is.
5. Depending on which configuration P has chosen, the following occurs:
   (a) if N is accepted as a child of P, P sends a WELCOME message to N, which N acknowledges immediatly (figures 1.(d) and 1.(e)). The join procedure is then completed for N.

(b) if either N or any of P's children (say C_j) is to be redirected to one of P's children (say C_k), that node is sent a GO(C_k) message (figure 1.(f)) and starts a join procedure (stage 1) with C_k assuming the role of P and C_j the one of N. When N receives such a message, it acknowledges this immediately with a GO_ACK message (figure 1.(g)). However, in order not to disrupt the flow of data for an already established node, C_j is given a time T_1 to find its new place in the tree.

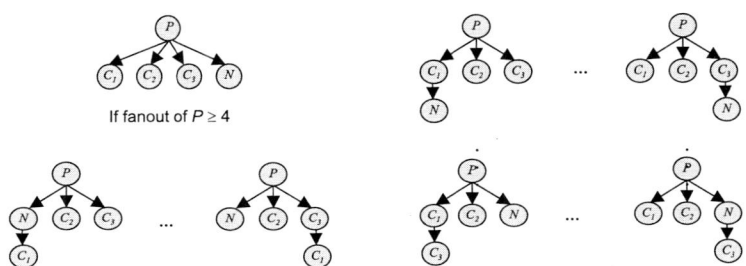

Fig. 2. Local Configurations Tested.

Notice that a Join procedure is not exclusively performed by TBCP entities that have not joined the tree yet. But, even a TBCP Entity that has already joined the tree may be forced to start a Join procedure, to find a new place in the tree. It should also be noted that the algorithm always finishes, as GO messages always send a TBCP entity (and the associated TBCP subtree whose root is the corresponding TBCP entity) *down* the TBCP tree.

Additional Rules for Tree Construction. In order to improve the efficiency (and the shape) of the control tree, a hierarchical organization of receivers in "domains" is also enforced, so that receivers belonging to the same domain belong to the same sub-tree.

Receivers declare their *domainID* when connecting to a new candidate parent (the *domainID* is a 32 bits identifier, e.g. IP address && netmask). The tree root can then elect a *domain root* node for each domain. For instance, the first node joining from a given domain can be elected as the domain root of its domain. Domain root nodes find their place in the tree with the same mechanism described in the previous section, starting from the tree root.

When a new node wants to join the tree, it is immediately redirected by the tree root to its domain's domain root with a GO message. The Join procedure described in the previous section then starts when the node sends the HELLO message to its domain root.

The following 2 constraints are also enforced, to keep all the nodes of the same domain "clustered":

1. Rule 1: A node P will discard configurations in which a node from its own domain is a child of a node from a different domain.
2. Rule 2: To keep domain roots as high as possible in the tree (i.e. as close as possible to the tree root), configurations in which a node P keeps more than one node from its own domain as children, and sends a domain root of a different domain as a child of one of its children, are discarded.

Figure 3 illustrates discarded configurations, where domains are represented by "colour codes".

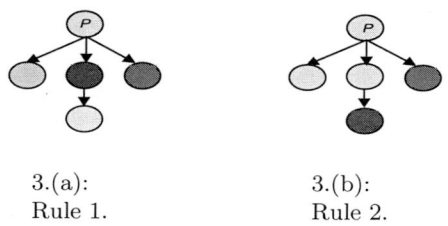

3.(a):
Rule 1.

3.(b):
Rule 2.

Fig. 3. Discarded Configuration Due to Rule Violation.

3.2   Cost Function and Measurements

In section 3.1, we have seen that a score is computed for each "local" configuration involving a parent node, its children and a "newcomer". These scores are obtained by applying to the local configurations a score function whose inputs are the distances among the nodes involved and/or any other relevant "metric" (e.g. domainIDs, node fanouts, etc.). Because the score function is used in each local decision taken by the TBCP nodes in the tree, the score function influences the final (i.e. global) shape of the tree. Of course, what constitutes a good shape for an overlay tree depends on the purpose for which the tree is built. Likewise, the notion of "distance" between tree nodes, used as input to the score function, can be represented by any metric (or set of metrics) of interest (e.g. delay, throughput, etc.) depending on the problem at hand.

Since score functions and distance evaluation mechanisms depend on the purpose of the tree, for flexibility, they are considered as "modules" rather than part of the core TBCP protocol.

For example, for an overlay used for reliable multicast, the following could be selected:

– distance $D(i, j)$ = round trip time (RTT) between node i and node j *along the tree*[2];

---

[2] Hence, if node i is the parent of node k that, in turn, is the parent of node j, then $D(i, j) = D(i, k) + D(k, j)$.

- score function = $max_{\forall M \in \{C_i\} \cup N} D(P, M)$, where $\{C_i\}$ is the set of P's children and N the newcomer.

The chosen configuration is the one with the smallest score. We therefore see that our strategy is to try and maximise the "control responsiveness" between nodes.

4   Simulation Results

We have simulated (with NS-2 [12]) our algorithm, with the score function proposed in section 3.2, in order to evaluate their suitability for purpose. At this stage, because we are mainly concerned with the characteristics of our trees, no background traffic is used and hop counts can therefore be used as distance metric instead of RTTs.

We have used topologies composed of:

- a core network of 25 routers, with each router connected to any other with a random probability of 10%;
- 15 stub networks of 10 routers each, with each router connected to any other router in the same stub with a random probability of 5%;
- each stub network is connected to the core by a single link, with each stub connected to a different core router.

Furthermore, the nodes running our algorithm (root/receivers) are "hosts" connected to stub routers only, with at most one receiver host connected to one stub router. The hosts are distributed randomly in the stub networks which determine domains. We therefore see that these topologies correspond to a worst-case scenario, as the distance between receivers and between stub networks is minimum 3 hops and there is no tendency to "cluster" the participating hosts. The results obtained should therefore indicate upper-bounds results.

We have simulated scenarios where all the TBCP entities have an identical fanout of respectively 2,3,4 and 5.

For each value of the fanout, groups of 3, 5, 10, 20, 30, 40, 50, 60, 70, 80, 90, 100 and 150 receivers have been incrementally tested on a same topology, that is 3 receivers joined the tree, then 2 were added, then 10, etc. This scenario was repeated 10 times (with a different topology each time) for each fanout value.

Since, in the context of this paper, we are interested in the performance of the TBCP trees for control purposes, the mean and maximum distance measured between any receiver (i.e. node) in the tree and its parent, as well as the distance between any receiver and its domain root (see section 3.1), are of prime concern. These are depicted in figure 4. In figure 4.(a), we see that as the size of the group increases, the mean distance between nodes and their parents decreases. This is because the larger the group is, the more "clustering" appears, and the algorithm is thus efficient at exploiting such clustering among receivers. The maximum distance observed between a node and its parent is due to the "interconnection" of different domains and depends on the topology.

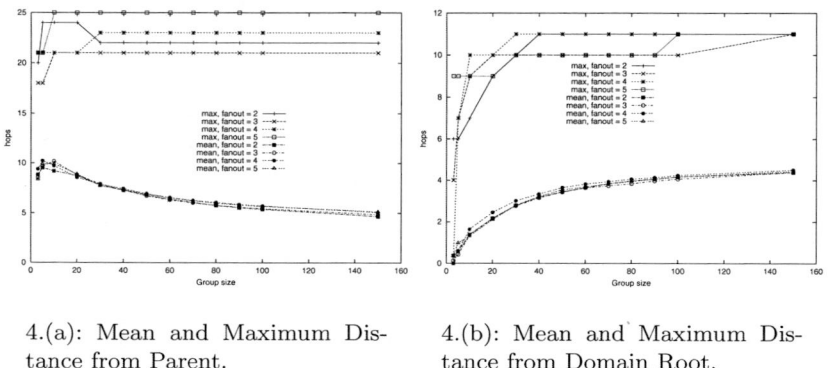

4.(a): Mean and Maximum Distance from Parent.

4.(b): Mean and Maximum Distance from Domain Root.

Fig. 4. Distances Measuring the "Locality" Efficiency of the TBCP Trees.

Figure 4.(b) shows that control operations confined within a domain would see small response times, with a mean distance between any node in a domain and its domain root (the root of the subtree covering the domain) increasing with a small slope. The position of the domain root within the domain has of course a great influence on the node-domain root distances, especially the maximum distance. Because in these simulations, the domain root was chosen to be the first node of a domain to join the tree and was therefore randomly placed within a domain, the maximum values in figure 4.(b) are therefore likely to represent worst-case scenarios.

The mean and maximum distances measured between any receiver and the tree root is depicted in figure 5. This figure represents the performance observed

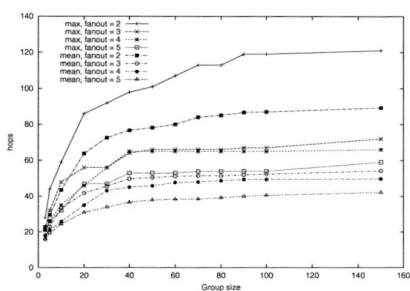

Fig. 5. Mean and Maximum Distance from (Tree) Root.

when traversing the tree from root to leaves. The distances between the root and

the leaf nodes is less critical in the control scenario considered here than if the
overlay tree were to be used for data transfers. As we expect, the values observed
stress the importance of the value of the fanout in reducing overall delay along
the tree, with the mean delay for a fanout of 2 being (roughly) 1.5 times the
maximum delay for a fanout of 5.

The tree can also be "globally" characterized by investigating mean and
maximum delay ratios and link stresses. The delay ratio for a receiver is defined
as the ratio of (the delay from the root to this receiver along the tree) to (the
delay from the root to this receiver with "direct" (unicast) routing). The link
stress is simply a count of how many replicates of the same data appear on a
link, when that data is "multicast" along the TBCP tree. These are depicted in
figures 6. Figure 6.(a) shows that, on average, the distance between the tree root

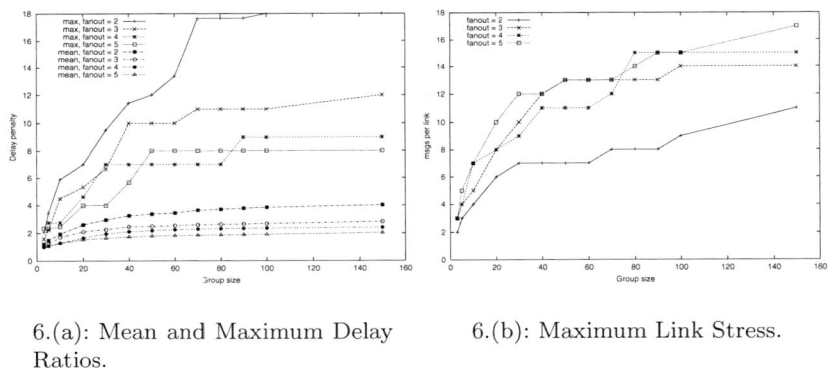

6.(a): Mean and Maximum Delay   6.(b): Maximum Link Stress.
Ratios.

Fig. 6. Global Characteristics of Tree.

and any node along the tree is on average about 2 to 4 times the direct distance
between the same nodes, which shows that the algorithm, and score function,
exhibits a tendency to grow the trees "in width". The figure also shows that
the value of the fanout has a dramatic effect on the maximum delay penalty
observed on the trees.

Figure 6.(b) shows the expected reduction in load on network links when an
overlay tree is used as opposed to a "reflector" which uses unicast communications from the root to each receiver. Indeed, in the reflector case, the maximum
link stress is equal to the group size, as the link directly out of the root must
carry all the data packets.

Finally, figures 7.(a) and 7.(b) illustrate the measurement sampling overhead
of the protocol (in terms of the number of nodes sampled) during the *initial*
joining period for a newcomer (i.e. from the moment the newcomer issues a JOIN
and it receives a WELCOME message). The overhead measured is therefore

proportional to the joining latency of a new node. Figure 7.(a) shows that the

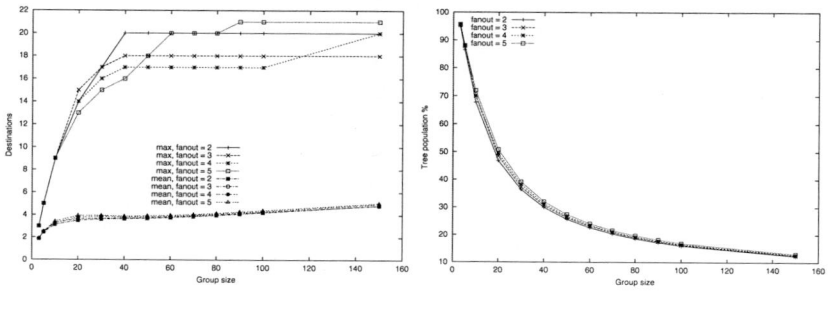

7.(a): Mean and Maximum Measurement Samples.

7.(b): Mean Measurement Samples in % of Population.

Fig. 7. Measurement Sampling Overhead

maximum number of nodes sampled is kept well below the number of nodes in the tree and is, indeed, very reasonable. It also shows that the average number of measurement samples taken by newcomers shows a sub-linear increase in terms of the group size.

Figure 7.(b) shows the percentage of the already existing population of the tree that is sampled by a newcomer. The shape of these curves should not be any surprise. Indeed, only nodes that are siblings of the nodes constituting the branch between the root and the final place of the newcomer are sampled in addition to nodes in this branch. Therefore, the more branches that are added to the tree, the bigger the proportion of the tree population will be ignored by each move of the newcomer along its branch.

These results presented in figure 7 show the scalability of the proposed tree building mechanism.

5   Conclusions

We have described an efficient and scalable TBCP algorithm/protocol to build such an overlay spanning tree. Our TBCP protocol does not rely on any special support from routers, nor on any knowledge of the network topology.

Because our algorithm builds the tree through a series of local decisions involving only a small subset of nodes for each decision, a joining node need only to get to know a few members of the multicast groups. Also, because our algorithm is decentralised, several new members can be in the process of joining the tree simultaneously without incurring additional overhead, which is a good scaling property for this type of protocols.

In the light of our simulation results, we believe our proposal constitutes a good candidate for scalable overlay tree building. We have also designed a leave procedure which, because of the lack of space, was not presented in this paper.

References

1. Y-H. Chu, S. Rao, and H. Zhang. A Case for End System Multicast. In *ACM SIGMETRICS 2000*, pages 1–12, Santa Clare, CA, USA, June 2000.
2. S. Deering and D. Cheriton. Multicast Routing in Datagram Internetworks and Extended LANs. *ACM Trans. on Comp. Syst.*, 8(2):85–110, May 1990.
3. C. Diot, B. Levine, B. Lyles, H. Kassem, and D. Balensiefen. Deployment Issues for the IP Multicast Service and Architecture. *IEEE Network*, 14(1):78–88, Jan/Feb 2000.
4. D. Farinacci, A. Lin, T. Speakman, and A. Tweedly. Pretty Good Multicast (PGM) Transport Protocol Specification. Internet Draft draft-speakman-pgm-spec-00, IETF, Jan 1998. Work in progress.
5. B. Fenner, M. Handley, H. Holbrook, and I. Kouvelas. Protocol Iindependent Multicast - Sparse Mode: Protocol Specification (revised). Internet Draft draft-ietf-pim-sm-v2-new-02, IETF, Mar 2001. Work in Progress.
6. S. Floyd, V. Jacobson, C. Liu, S. McCanne, and L. Zhang. A Reliable Multicast Framework for Light-weight Sessions and Application Level Framing. *IEEE/ACM Trans. Network.*, 5(6):784–803, Dec 1997.
7. P. Francis. Yoid: Extending the Internet Multicast Architecture. Technical report, ACIRI, Apr 2000. http://www.aciri.org/yoid.
8. H. Holbrook and D. Cheriton. IP Multicast Channels: EXPRESS Support for Large-scale Single-source Applications. *ACM Comp. Comm. Reviews*, 29(4):65–78, Oct 1999.
9. J. Jannotti, D. Gifford, K. Johnson, F. Kaashoek, and J. O'Toole. Overcast: Reliable Multicasting with an Overlay Network. In *USENIX OSDI 2000*, San Diego, CA, USA, Oct 2000.
10. M. Kadansky, D. Chiu, and J. Wesley. Tree-based Reliable Multicast (TRAM). Internet Draft draft-kadansky-tram-00, IETF, Nov 1998. Work in Progress.
11. B. Levine and JJ. Garcia-Luna. Improving Internet Multicast with Routing Labels. In *IEEE Intl. Conf. on Network Protocols (ICNP'97)*, pages 241–250, Atlanta, USA, Oct 1997.
12. Ns-2 Network Simulator. http://www.isi.edu/nsnam/ns/.
13. D. Pendarakis, S. Shi, D. Verma, and M. Waldvogel. ALMI: an Application Level Multicast Infrastructure. In *3rd USENIX Symposium on Internet Technologies*, San Francisco, CA, USA, Mar 2001.
14. B. Whetten, D.M. Chiu, M. Kadansky, and G. Taskale. Reliable Multicast Transport Building Block for TRACK. Internet Draft draft-ietf-rmt-bb-track-01, IETF, Mar 2001. Work in progress.
15. B. Whetten and G. Taskale. An Overview of Reliable Multicast Transport Protocol II. *IEEE Network*, 14(1):37–47, Jan 2000.
16. R. Yavatkar, J. Griffioen, and M. Sudan. A Reliable Dissemination Protocol for Interactive Collaborative Applications. In *ACM Multimedia'95*, San Francisco, CA, USA, Nov 1995.

The Multicast Bandwidth Advantage in Serving a Web Site

Yossi Azar[1], Meir Feder[2], Eyal Lubetzky[3],
Doron Rajwan[3], and Nadav Shulman[3]

[1] Dept. of Computer Science, Tel-Aviv University, Tel-Aviv, 69978, Israel
azar@tau.ac.il
[2] Bandwiz, Israel and Department of Electrical Engineering - Systems
Tel Aviv University, Tel-Aviv, 69978, Israel Meir@bandwiz.com
[3] Bandwiz, Israel
{EyalL,Doron,Nadav}@bandwiz.com

Abstract. Delivering popular web pages to the clients results in high
bandwidth and high load on the web servers. A method to overcome this
problem is to send these pages, requested by many users, via multicast.
In this paper, we provide an analytic criterion to determine which pages
to multicast, and analyze the overall saving factor as compared with
a unicast delivery. The analysis is based on the well known observation
that page popularity follows a Zipf-like distribution. Interestingly, we can
obtain closed-form analytical expressions for the saving factor, that show
the multicast advantage as a function of the site hit-rate, the allowed
latency and the Zipf parameter.

1 Introduction

One of the largest problems in the web is to deliver the content efficiently from
the site to the user. High load on the server and on the network leads to long
delays or more extremely denial of services. Increasing the capacity for delivering
the content results in a high cost of extra servers and extra bandwidth. Moreover,
the capacity is planed to some value, though larger than the average load, but
almost always cannot accommodate the peak load. This is specially correct for
popular pages were the access pattern may be unpredictable and very unstable
(e.g. the famous Starr report case).

There are several methods to try to overcome the problem. One is to use
caches [13, 7, 1]. However, caches are not effective for frequently changing content
or for long files (e.g video, audio). A different possibility that we consider in this
paper is to use multicast [4, 8, 6], i.e., to deliver the content simultaneously to
many (all) users via multicast dynamic tree. Obviously, one may also combine
both caching and multicasting to further improve the solution.

At first, it may seem that multicast could be effective only if many users
requests exactly the same content at exactly the same time, which can occur
mainly in real time events. However, it is well known (see, e.g., [8]) that one
can cyclicly transmit by multicast a page until all users requested the page in

J. Crowcroft and M. Hofmann (Eds.): NGC 2001, LNCS 2233, pp. 88–99, 2001.

the multicast tree receive it. Note that each user needs to receive one cycle from the time that he joins the tree (which does not need to be a beginning of a new cycle) assuming that there are no faults. A more efficient methods that overcomes possible packet losses can be achieved by using erasure codes, e.g., [11, 3].

The multicast advantage is manifested by combining together overlap requests to a single transmission. This way the server load and bandwidth decrease dramatically since all overlapped users appear almost as a single user. Hence, the most attractive pages (files) to multicast are pages that are popular, i.e., have many hits per second, and pages that are large. Fortunately, the access pattern to pages of a site are far from being uniform. Any non-uniformity on the distribution of the access pattern to pages enhances the advantage of using multicast since it results in more popular, hence higher concurrency, pages. It has been observed [2, 10, 9] that indeed the access pattern for pages in a site is highly non-uniform and obeys a Zipf-like distribution with α parameter that is in the range of $1.4 - 1.6$. With this distribution, a fixed number of pages account for almost all requests for pages (say 95%). As in many other events, Zipf distribution occurs naturally, and so we assume that this is the request pattern in order to obtain quantitative expressions for the multicast advantage. We will present the results in terms of the Zipf parameter α and note that even for the pure Zipf distribution, i.e. for parameter $\alpha = 1$, and furthermore even for Zipf-like distribution with $\alpha < 1$, a small number of pages (maybe not as small as for $\alpha > 1$) still account for most of the requests. Since a Zipf-like distribution has a heavy tail, assuming such a distribution on the access pattern is one of the weakest possible assumptions in terms of the advantage of multicast.

It is worthwhile to mention that the popular pages may change over time. An appropriate system that keeps track of the access pattern can easily maintain the list of the hot pages. Hence, such a system can decide which pages to multicast at each point in time according to the estimated parameters of the access rate and the size of the pages.

We next discuss the results of this paper. We start, in section 2, by an analysis of a site in which all the information regarding the access pattern and file distribution is given. The analysis is based on a criterion we derive, that determines which pages to multicast. This criterion assumes that the page access rate is given, or estimated, and it also depends on the allowable delay to receive the page, which in turn, determines the bandwidth in which the page is multicasted. The major result of our paper appears in section 3, and contains a set of analytical expression for the gain in bandwidth (and server load) in serving a *typical site* by selective multicast (i.e., multicast of hot pages) as compared with the standard unicast serving. For the typical site we assume that the access pattern follows a Zipf-like distribution with some parameter α. The overall saving bandwiz factor achieved depends on the access rate to the site and the latency that we allow for pages. Section 4 extends the analysis to a site with various typical file groups. The paper is summarized in section 5.

2   Analysis for a Given Site

We make the following notations

- n the number of pages in the site.
- p_i probability of requesting page i for $1 \leq i \leq n$ given that a page was requested from the site.
- S_i is the size of page i, in bits, for $1 \leq i \leq n$.
- λ the average access rate in hits per unit time, to the site. We note that $\lambda = N\lambda_0$ where N is the size of the population accessing the site and λ_0 is the average access rate of a person from the population to the site.

As a step toward an analysis for a typical site we make an analysis for a given site with the probably unrealistic assumption that all the above parameters (n, p_i, S_i, λ) are known. In this section we first compute the minimal required bandwidth to serve this site by unicast. We then consider serving the site by selective multicast, where we first determine which pages worth multicasting and then compute the resulting bandwidth. By that we estimate the gain in serving this site by multicast. Note that we assume that the site is planned to have the ability of serving all requests and not to drop/block some of them.

2.1   Serving by Unicast

Using the above notation the amount of bits per unit time generated on the average in serving the page i is $\lambda p_i S_i$. Consider now

$$B_u = \sum_{i=1}^{n} \lambda p_i S_i = \lambda \sum_{i=1}^{n} p_i S_i \ .$$

This formula is the information theoretic lower bound on the required bandwidth for serving all the pages by unicast, since the total average number of bits requested per unit time must by equal (on the average) to the total number of bits transmitted. Note that the lower bound is independent of the transmission rate of the pages. Moreover, the above formula stands for the minimum possible bandwidth in the ideal case where we can store the requests in a queue and output continuously exactly the same number of bits per time without any bound on the latency encountered for delivering the files. The actual bandwidth required by any practical system to support all requests (in particular, with bounded latency) needs to be higher than this. Nevertheless, we essentially demonstrate the multicast bandwidth advantage by showing that multicast requires less (sometimes much less) bandwidth than this information theoretic bound.

2.2   Serving by Selective Multicast

In serving a file i by multicast, a carousel transmission (or better, a coded stream using, e.g., Bandwiz block-to-stream code [11]) of the file is transmitted at some particular bandwidth w_i and all requests for the file are handled by receiving from this multicast transmission. The bandwidth advantage in serving a file this way comes from the fact that the file is served at the fixed bandwidth w_i and this bandwidth allocation is sufficient no matter how many requests the file has during its transmission. In unicast, on the other hand, each request requires an additional bandwidth allocation.

One may conclude that multicast can lead to an unbounded saving compared with unicast, simply by allocating a small bandwidth w_i to serve the file i. But there is a price for that. The latency in receiving the file, whose size is S_i will become large. A reasonable multicast bandwidth allocation is such that the desired latency L_i is guaranteed. Note that the information theoretic lower bound computed for unicast was independent of the latency we allow to deliver any file (although the realistic bandwidth, higher than that, does depend on it as discussed above). Thus, as the allowed latency is larger, the multicast advantage is larger.

In view of this discussion, we assume that the bounds on the latencies for the various files are imposed on the system. We use the following definitions:

- Let L_i be the latency we allow for delivering page i using multicast.
- Thus, $w_i = S_i/L_i$ is the rate that we chose to transmit page i.

We note that the value of w_i and L_i are functions of the typical capability of the receivers and network conditions. For example, w_i should not be larger than the typical modem rate if typical receivers access the site through a modem. This implies that L_i cannot be small for large files. Also for small files it does not pay to have small L_i since creating the connection from the receiver to the site would dominate the delay. Hence we conclude that L_i is never very small and may be required to be reasonably large. As will be seen, the larger the L_I, the better is the multicast advantage.

Out of the bandwidth allocated to unicast, the portion of the minimal bandwidth required to transmit the file i is $\lambda p_i S_i$ (which is the amount of bits per unit time requested of this file). Thus, in using multicast, we reduce the bandwidth to all the pages in which

$$\lambda p_i S_i > w_i$$

and in this case we replace $\lambda p_i S_i$ by the bandwidth by w_i. The above formula, which provides the criterion for transmitting the file by multicast, is equivalent to

$$\lambda p_i L_i > 1 \ .$$

Hence we conclude that the total bandwidth required by the selective multicast is

$$B_m = \sum_{i \mid \lambda p_i L_i > 1} w_i + \sum_{i \mid \lambda p_i L_i \leq 1} \lambda p_i S_i \ .$$

3   Analysis for a Typical Site

We consider a site, where the various pages can be partitioned into typical groups. In each group the pages are of similar characteristics, i.e. approximately the same size and same required latency for delivery to the user. For example, one group can be text HTML files, another group can be pages with images and yet another group can be audio, or video files. We first consider one such group of pages. It is well known and has been consistently observed that the access pattern to the pages in a group is not uniform. In fact, the advantage of multicast improves as the distribution becomes less uniform since one needs to multicast less pages to deliver the same fraction of the traffic. We make one of the weakest possible assumptions on that distribution, i.e., a family of heavy tail distributions on the access pattern. If the distribution is more skewed then the saving by using multicast increases.

Assumption. Among a group of pages with the same latency the popularity of pages is distributed according to Zipf-like distribution with some parameter $\alpha > 0$. Specifically, the probability of the i'th most popular page is proportional to $1/i^\alpha$ or equal to $\frac{1}{C(\alpha)i^\alpha}$ where $C(\alpha) = \sum_{i=1}^{n} \frac{1}{i^\alpha}$.

The above assumption is crucial for our analysis. The typical parameter α which is usually observed for a typical site is in the range $1.4 - 1.6$. In the sequel we will use the following approximation $\sum_{i=a+1}^{b} \frac{1}{i^\alpha} \approx \int_a^b \frac{1}{x^\alpha} dx$ or $\sum_{i=a}^{b} \frac{1}{i^\alpha} \approx \frac{1}{a^\alpha} + \int_a^b \frac{1}{x^\alpha} dx$. In particular $\sum_{i=1}^{n} \frac{1}{i^\alpha} \approx 1 + \int_1^n \frac{1}{x^\alpha} dx$.

Now, we are ready to continue the analysis. First we consider unicast. We can approximate the expression

$$B_u = \lambda \sum_{i=1}^{n} p_i S_i$$

by

$$B_u = \lambda E(S)$$

where $E(S)$ is the expected size of a random page in the group.

Using the Zipf-like distribution we can evaluate the total bandwidth required by multicast. Recall that it is worthwhile to multicast a page if $\lambda p_i L > 1$ (L is fixed for all pages in the group) and we should multicast the most popular pages regardless of their size. Let k be the number of such pages that are worth to multicast. Then k is the largest integer that satisfies $\lambda p_k L > 1$ or

$$\frac{1}{C(\alpha)k^\alpha} = p_k \geq \frac{1}{\lambda L} .$$

Following the above formula there are three different cases that we need to analyze according the to value of the smallest k that satisfies the above formula:

- No need to multicast any page. This is the case where the access rate is small and the required latency is so short that it is not worthwhile to multicast even the most popular page (smallest $k \leq 1$). That corresponds to $\lambda L \leq C(\alpha)$.

- Multicast all pages. Here the access rates are high or the number of pages is relatively small such that it is worthwhile to multicast all pages ($k \geq n$). Here all pages are popular which corresponds to $\lambda L \geq C(\alpha)n^{\alpha}$.
- Multicast popular pages. This is the typical case where $1 < k < n$ and we multicast only the popular pages according to our metric. This corresponds to $C(\alpha) < \lambda L < C(\alpha)n^{\alpha}$.

Clearly, in the first case multicast saves nothing. Later we discuss the saving when we multicast all pages. We begin, then, with the interesting case where $1 < k < n$, i.e., the case of multicasting only the popular pages.

3.1 Multicasting the Popular Pages

In this case we get $k = \left\lfloor \left(\frac{\lambda L}{C(\alpha)} \right)^{1/\alpha} \right\rfloor$ where $1 \leq k \leq n$.

If we plug it into the formula of the total bandwidth of the multicast (i.e. multicast the first k pages and unicast the rest) we get

$$B_m = \sum_{i=1}^{k} S_i/L + \sum_{i=k+1}^{n} \lambda p_i S_i .$$

Since the pages in a group have similar characteristics in terms of size and required latency we can approximate the above by the following

$$B_m \approx \frac{E(S)}{L} \left(\frac{\lambda L}{C} \right)^{1/\alpha} + \frac{E(S)\lambda}{C} \int_{(\frac{\lambda L}{C})^{1/\alpha}}^{n} \frac{1}{x^{\alpha}} dx$$

$$= \frac{E(S)\lambda}{C} \left(\left(\frac{\lambda L}{C} \right)^{1/\alpha-1} + \int_{(\frac{\lambda L}{C})^{1/\alpha}}^{n} \frac{1}{x^{\alpha}} dx \right)$$

where we drop the integer value and we set

$$C = C(\alpha) = 1 + \int_{1}^{n} \frac{1}{x^{\alpha}} dx .$$

Next we separate between the case $\alpha = 1$ and the case $\alpha \neq 1$. For the case $\alpha \neq 1$ we also consider asymptotic behavior.

the Case $\alpha = 1$. Clearly

$$C = 1 + \int_{1}^{n} \frac{dx}{x} = 1 + \ln n - \ln 1 = \ln en$$

and

$$\int_{\frac{\lambda L}{C}}^{n} \frac{dx}{x} = \ln n - \ln \frac{\lambda L}{C} = \ln \frac{nC}{\lambda L} = \ln \frac{n \ln en}{\lambda L} .$$

Hence for the range of the typical case i.e., $\ln en < \lambda L < n \ln en$, we have

$$B_m \approx \frac{E(S)\lambda}{\ln en}\left(1 + \ln\frac{n\ln en}{\lambda L}\right) = E(S)\lambda\left(\frac{\ln n + 1 + \ln\frac{\ln en}{\lambda L}}{\ln en}\right)$$

$$= E(S)\lambda\left(\frac{\ln en - \ln\frac{\lambda L}{\ln en}}{\ln en}\right) = E(S)\lambda\left(1 - \frac{\ln\frac{\lambda L}{\ln en}}{\ln en}\right).$$

If we compare it to standard unicast, the saving factor is

$$R = \frac{1}{1 - \frac{\ln\frac{\lambda L}{\ln en}}{\ln en}}.$$

Examples of the savings can be seen in Table 1. Here λ is given in hits per second for the site (i.e. total rate for all pages), L is given is seconds (4 seconds for html page, 20 seconds for page with pictures and 300 seconds for audio or video clip) and n is the number of pages of the site. Plots of R appear in Figure 2 as a function of λ (and also for various α's, see also below).

λ	L	n	saving, $\alpha = 1$
200	20	10^4	2.41
200	4	10^3	2.40
20	300	10^3	6.19

Fig. 1. Examples of the Saving Factor for $\alpha = 1$.

the Case $\alpha \neq 1$. In this case

$$C = 1 + \int_1^n \frac{dx}{x^\alpha} = 1 + \frac{n^{1-\alpha} - 1}{1 - \alpha} = \frac{n^{1-\alpha} - \alpha}{1 - \alpha}$$

and

$$\int_{(\frac{\lambda L}{C})^{1/\alpha}}^n \frac{dx}{x^\alpha} = \frac{n^{1-\alpha} - (\frac{\lambda L}{C})^{\frac{1-\alpha}{\alpha}}}{1 - \alpha}.$$

Hence for the range

$$\frac{n^{1-\alpha} - \alpha}{1 - \alpha} < \lambda L < \frac{n^\alpha(n^{1-\alpha} - \alpha)}{1 - \alpha}$$

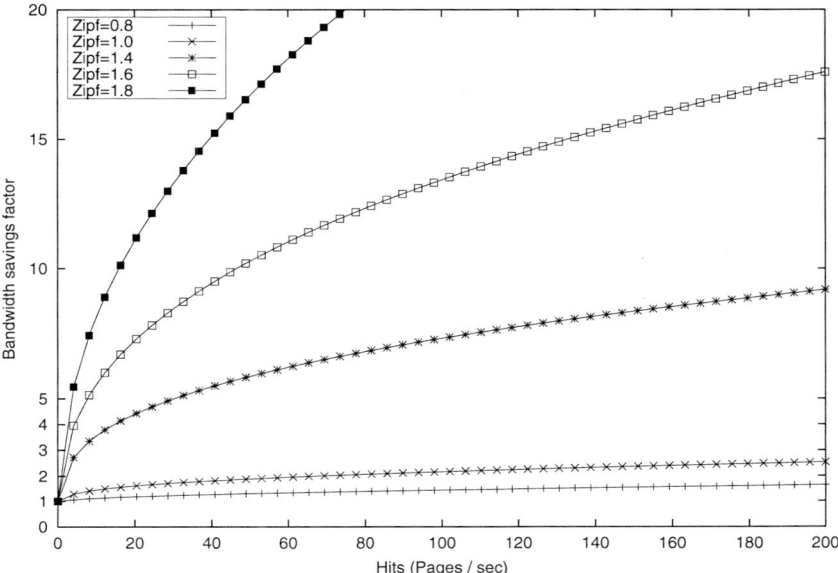

Fig. 2. The saving factor (relative to unicast) of the bandwidth (load) of a server for multicast with Zipf-like distribution for various values of the parameter α as a function of the number of hits per second. The number of pages is $10,000$ and the latency is 25 seconds.

we have

$$
\begin{aligned}
B_m &\approx \frac{E(S)\lambda}{C}\left(\left(\frac{\lambda L}{C}\right)^{1/\alpha-1} + \frac{n^{1-\alpha} - (\frac{\lambda L}{C})^{1/\alpha-1}}{1-\alpha}\right)\\
&= \frac{E(S)\lambda}{C}\left(\left(\frac{\lambda L}{C}\right)^{1/\alpha-1}(1 - \frac{1}{1-\alpha}) + \frac{n^{1-\alpha}}{1-\alpha}\right)\\
&= \frac{E(S)\lambda}{(1-\alpha)C}\left(-\alpha\left(\frac{\lambda L}{C}\right)^{1/\alpha-1} + n^{1-\alpha}\right)\\
&= \frac{E(S)\lambda}{n^{1-\alpha}-\alpha}\left(-\alpha\left(\frac{\lambda L(1-\alpha)}{n^{1-\alpha}-\alpha}\right)^{1/\alpha-1} + n^{1-\alpha}\right)\; .
\end{aligned}
$$

We conclude that the saving factor compared with unicast is

$$
R = \frac{n^{1-\alpha}-\alpha}{n^{1-\alpha}-\alpha\left(\frac{\lambda L(1-\alpha)}{n^{1-\alpha}-\alpha}\right)^{1/\alpha-1}}\; .
$$

Again, plots of R as a function of λ and various α's appear in Figure 2.

Asymptotic Expression - $\alpha > 1$. It is interesting to consider the asymptotic behavior of the saving factor for a site, as the number of pages grows. It is not hard to show that the saving function is monotone non increasing with the number of pages. Moreover, for the case $\alpha > 1$, it turns out that the saving factor approaches to a limit which is bounded away from 1. Hence, to bound the saving factor for any number of pages we can assume that the number of pages n approaching infinity. The saving factor R in the asymptotic case, which as will be seen has a simpler expression (independent of n), is a lower bound on the saving factor for any n (i.e. we save at least that much). This is very useful since the number of pages in a site is usually large and continuously growing.

For evaluating the asymptotic behavior we approximate the expression for R by replacing $n^{1-\alpha}$ with zero. Then for the range $\frac{\alpha}{\alpha-1} < \lambda L$ we have

$$B_m \approx \frac{E(S)\lambda}{-\alpha}\left(-\alpha\left(\frac{\lambda L(1-\alpha)}{-\alpha}\right)^{1/\alpha-1}\right)$$
$$= E(S)\lambda\left(\lambda L(1-1/\alpha)\right)^{1/\alpha-1} .$$

Hence the saving factor relative to unicast is

$$R = \left(\lambda L(1-1/\alpha)\right)^{1-1/\alpha}$$

and it is independent of n.

The saving factor of the total bandwidth for a site (including both unicast pages and multicast pages) yields by multicasting the relevant pages can be found in Figure 3 for $\alpha = 1.4$, $\alpha = 1.6$ and $\alpha = 1.8$ for few examples.

λ	L	saving, $\alpha = 1.4$	saving, $\alpha = 1.6$	saving, $\alpha = 1.8$
200	20	7.48	15.25	27.82
200	4	4.72	8.49	13.60
20	300	8.39	18.07	33.31

Fig. 3. Examples of the Saving Factor for $\alpha = 1.4$, $\alpha = 1.6$ and $\alpha = 1.8$.

Asymptotic Expression - $\alpha < 1$. Now assume that $\alpha < 1$. For the asymptotic behavior we can approximate the expression by assuming that $n^{1-\alpha}$ is relatively large compare to α (i.e n is relatively large). Then for the approximate range

$$\frac{n^{1-\alpha}}{1-\alpha} < \lambda L < \frac{n}{1-\alpha}$$

we have

$$B_m \approx \frac{E(S)\lambda}{n^{1-\alpha}} \left(-\alpha \left(\frac{\lambda L(1-\alpha)}{n^{1-\alpha}} \right)^{1/\alpha-1} + n^{1-\alpha} \right)$$

$$= E(S)\lambda \left(1 - \frac{\alpha}{n^{1-\alpha}} \left(\frac{\lambda L(1-\alpha)}{n^{1-\alpha}} \right)^{1/\alpha-1} \right)$$

$$= E(S)\lambda \left(1 - \alpha \left(\lambda L(1-\alpha)/n \right)^{1/\alpha-1} \right) .$$

Hence the saving factor is

$$R = \frac{1}{1 - \alpha \left(\lambda L(1-\alpha)/n \right)^{1/\alpha-1}}$$

relative to unicast. This expression depends on n (as n goes to infinity, the saving factor goes to 1, i.e., no saving) but it is a simpler expression than above.

3.2 Multicast All Pages

Here we multicast all pages i.e., $k = n$ which corresponds to the range $\lambda L \geq C(\alpha)n^\alpha$. We have $B_m = \sum_{i=1}^{n} S_i/L = E(S)n/L$. If we compare it to unicast, we get that the saving factor is

$$R = \frac{\lambda L}{n} .$$

It is worthwhile to note that the above saving factor holds for all values of α. The range for achieving this saving factor is $\lambda L \geq n \ln en$ for $\alpha = 1$ and $\lambda L \geq \frac{n^\alpha (n^{1-\alpha} - \alpha)}{1-\alpha}$ for $\alpha \neq 1$. The range for $\alpha \neq 1$ can be approximated by the range $\lambda L \geq \frac{\alpha n^\alpha}{1-\alpha}$ for $\alpha > 1$ and $\lambda L \geq \frac{n}{1-\alpha}$ for $\alpha < 1$.

It is also worthwhile to mention that the case $\alpha = 0$ (i.e. uniform distribution) always falls in the extreme case or the low traffic. That is if $\lambda L > n$ it is worth while to multicast all pages and otherwise it is not worthwhile to multicast any page.

3.3 Properties of the Saving Function

We list the following useful observations:

- The saving function is continuous monotone non-decreasing as a function of λL for any given α and n in the admissible range. This can be easily proved by considering the saving function directly.
- The saving function is continuous monotone non-increasing as a function of n for any given α and λL in the admissible range. This can be easily proved for $\alpha = 1$. For $\alpha \neq 1$ this can be proved by showing that the saving function is monotone in $n^\alpha - \alpha$ which is monotone in n.
- The saving function seems to be continuous monotone non-decreasing as a function of α (also at $\alpha = 1$) for any given n and λL in the admissible range.

4   A Site with Various Groups

In this section we assume that not all pages have a similar size and latency. We partition the files into r groups where in each group the files are of approximately the same size and latency. For group i we denote by $f_u^j(E_j(S), \lambda_j))$ the average bandwidth required for group j using the standard unicast serving and by $f_m^j(E_j(S), \lambda_j, L_j)$ the average bandwidth required for group j using multicast. Recall that we do not limit the number of pages that we multicast and hence, the decision if to multicast a page does not conflict with the decisions to multicast other pages. Hence the overall bandwidth is superposition of the bandwidth of the individual groups. Thus, we have that the total bandwidth used in unicast is

$$\sum_{j=1}^{r} f_u^j(E_j(S), \lambda_j))$$

where $f_u^j(E_j(S), \lambda_j)) = \lambda_j E_j(S)$. The total bandwidth for multicast serving is

$$\sum_{j=1}^{r} f_m^j(E_j(S), \lambda_j L_j)$$

where for group j of the extreme case

$$f_m^j(E_j(S), \lambda_j, L_j) = E_j(S)n/L_j$$

and for group j of the typical case with $\alpha = 1$

$$f_m^j(E_j(S), \lambda_j, L_j) = E_j(S)\lambda_j \left(1 - \frac{\ln \frac{\lambda L_j}{\ln en_j}}{\ln en_j} \right)$$

where for $\alpha \neq 1$

$$f_m^j(E_j(S), \lambda_j, L_j) = \frac{E_j(S)\lambda_j}{n_j^{1-\alpha_j} - \alpha_j} \left(-\alpha_j \left(\frac{\lambda_j L_j(1 - \alpha_j)}{n_j^{1-\alpha_j} - \alpha_j} \right)^{1/\alpha_j - 1} + n_j^{1-\alpha_j} \right).$$

5   Summary

Our main contribution in this paper is the analytical analysis of the saving factor that can be achieved by using multicast versus using unicast in serving a a typical site. The analysis assumes the Zipf-like distribution for the access pattern for the pages in the site. We note that for the most interesting case where the parameter α of the Zipf-like distribution is larger than 1 the saving factor is almost independent of the number of pages (i.e the site may contain a huge number of pages). We also note that a crucial parameter in determining the saving factor is the product between λ and L which is the access rate for a group and the maximum latency we are allowed to deliver the files. We have also designed a simple criterion for a given site to decide in advance (or dynamically while collecting the information on the access pattern for the site) which pages to multicast and which pages to continue to transmit with the standard unicast.

We note that the saving factor can be further improved, if we further consider the peak behavior and not the average behavior of the requests. In this case the requirement for unicast bandwidth grow, while the requirement for multicast is stable. We can change somewhat the criterion of which pages to multicast - instead of comparing the average required rate for sending a page in unicast to its multicast bandwidth, we compare the instantaneous demand. The exact analysis in this case requires assumptions regarding the stochastic access pattern. Recent studies show that requests are not coming as, say, a Poisson process, but have a self-similar heavy tail distribution (see e.g. [12, 5]). Thus, this analysis can be complicated. Still, an approximation for the true saving can be obtained by using the results derived here, and choosing for λ a higher value, that will reflect the peak demand instead of the average access rate.

References

[1] J. Angel. Caching in with content delivery. *NPN: New Public Network Magazine*, http://www.networkmagazine.com, 2000.

[2] L. Breslau, P. Cao, L. Fan, G. Phillips, and S. Shenker. Web caching and zipf-like distributions: Evidence and implications. In *Infocom*, 1999.

[3] John W. Byers, Michael Luby, Michael Mitzenmacher, and Ashutosh Rege. A digital fountain approach to reliable distribution of bulk data. In *SIGCOMM*, pages 56–67, 1998.

[4] R. J. Clark and M. H. Ammar. Providing scalable Web services using multicast communication. *Computer Networks and ISDN Systems*, 29(7):841–858, 1997.

[5] M. Crovella and A. Bestavros. Self-similarity in World Wide Web traffic: Evidence and possible causes. *IEEE/ACM Transactions on Networking*, 5(6):835–846, 1997.

[6] D. Dolev, O. Mokryn, Y. Shavitt, and I. Sukhov. An integrated architecture for the scalable delivery of semi-dynamic web content. Technical report, Computer Science, Hebrew University, 2000.

[7] A. Dornan. Farming out the web servers. *NPN: New Public Network Magazine*, http://www.networkmagazine.com, 2000.

[8] Z. Fei, K. Almeroth, and M. Ammar. Scalable delivery of web pages using cyclic-best-effort (udp) multicast. In *Infocom*, 1998.

[9] P. Krishnan, D. Raz, and Y. Shavitt. The cache location problem. In *IEEE/ACM Transaction on Networking (ToN)*, 2000.

[10] V. Padmanabhan and L. Qiu. The content and access dynamics of a busy web site: Findings and implications. In *ACM SIGCOMM 2000*, 2000.

[11] Bandwiz White Paper. http://www.bandwiz.com/solu_library.htm.

[12] V. Paxson and S. Floyd. Wide area traffic: the failure of Poisson modeling. *IEEE/ACM Transactions on Networking*, 3(3):226–244, 1995.

[13] A. Wolman, G. M. Voelker, N. Sharma, N. Cardwell, A. R. Karlin, and H. M. Levy. On the scale and performance of cooperative web proxy caching. In *Symposium on Operating Systems Principles*, pages 16–31, 1999.

STAIR: Practical AIMD Multirate Multicast Congestion Control

John Byers and Gu-In Kwon[*]

Computer Science Department
Boston University, Boston, MA 02215
{byers,guin}@cs.bu.edu

Abstract. Existing approaches for multirate multicast congestion control are either friendly to TCP only over large time scales or introduce unfortunate side effects, such as significant control traffic, wasted bandwidth, or the need for modifications to existing routers. We advocate a layered multicast approach in which steady-state receiver reception rates emulate the classical TCP sawtooth derived from additive-increase, multiplicative decrease (AIMD) principles. Our approach introduces the concept of dynamic *stair* layers to simulate various rates of additive increase for receivers with heterogeneous round-trip times (RTTs), facilitated by a minimal amount of IGMP control traffic. We employ a mix of cumulative and *non-cumulative* layering to minimize the amount of excess bandwidth consumed by receivers operating asynchronously behind a shared bottleneck. We integrate these techniques together into a congestion control scheme called STAIR which is amenable to those multicast applications which can make effective use of arbitrary and time-varying subscription levels.

1 Introduction

IP Multicast will ultimately facilitate both delivery of real-time multimedia streams and reliable delivery of rich content to very large audience sizes. One of the most significant remaining impediments to widespread multicast deployment is the issue of congestion control. Internet service providers and backbone service providers need assurances that multicast traffic will not overwhelm their infrastructure. Conversely, content providers in the business of delivering content via multicast do not want artificial handicaps imposed by overly conservative multicast congestion control mechanisms. Resolution of the tensions imposed by this fundamental problem in networking motivates careful optimization of multicast congestion control algorithms and paradigms.

While TCP-friendly multicast congestion control schemes which transmit at a single rate now exist, these techniques cannot scale to large audience sizes. The apparent alternative is *multirate* congestion control, whereby different receivers in the same session can receive content at different transfer rates. Several schemes for multirate congestion control using layered multicast [1,7,8,11,13] have been

---

[*] Work supported in part by NSF Grants CAREER ANI-0093296 and ANI-9986397.

J. Crowcroft and M. Hofmann (Eds.): NGC 2001, LNCS 2233, pp. 100–112, 2001.

proposed. Also, an excellent survey of related work on multicast congestion control appears in [12]. A layered multicast approach employs multiple multicast groups transmitting at different rates to accommodate a large, heterogeneous population of receivers. In these protocols, receivers adapt their reception rate by subscribing to and unsubscribing from additional groups (or layers), typically leveraging the Internet Group Membership Protocol (IGMP) as the control mechanism. Also, these schemes tend to employ cumulative layering, which mandates that each receiver always subscribe to a set of layers in sequential order. Cumulative layering dovetails well with many applications, such as those which employ layered video codecs [8] for video transmission and methods for reliable multicast transport which are tolerant to frequent subscription changes [11,4].

In conventional layering schemes, the rates for layers are exponentially distributed: the base layer's transmission rate is B_0, and all other layers i transmit at rate $B_0 * 2^{i-1}$. Therefore, subscription to an additional layer doubles the receiver's reception rate. Reception rate increase granularity of those schemes is unlike TCP's fine-grained additive-increase, multiplicative decrease (AIMD). Because of this coarse granularity, rate increases are necessarily abrupt, which runs the risk of buffer overflow; therefore, receivers must carefully infer the available bandwidth before subscribing to additional layers.

A different approach advocating fine-grained multicast congestion control to simulate AIMD was proposed in [3]. We refer to this approach as FGLM (Fine-Grained Layered Multicast). FGLM relies on *non-cumulative* layering and careful organization of layer rates to enable a receiver to increase the reception rate at the granularity of the base layer bandwidth B_0. Unlike earlier schemes, in this scheme, all receivers act autonomously with no implicit or explicit coordination between them. One substantial drawback of this approach is a constant hum of IGMP traffic at each last hop router (1 join and 2 leaves per client at every additive increase decision point). This volume of control traffic is especially problematic for last hop routers with large fanout to one multicast session, or those serving multiple sessions. Another drawback is that this approach incurs some bandwidth *dilation* at links, wasted bandwidth introduced by the uncoordinated activities of the set of downstream receivers. Finally, the use of non-cumulative layers is only amenable to applications which can make use of an arbitrary (and frequently changing) subset of subscription layers over time. The most natural applications of which we are aware are those in which any packet on any layer is equivalently useful to every receiver; such a situation arises in the digital fountain approach defined in [4], which facilitates reliable multicast by transmitting content encoded with fast forward error correcting codes.

Our work presents a better method for simulating true AIMD multicast congestion control. At a high level, our STAIR (Simulate TCP's Additive Increase/multiplicative decrease with Rate-based) multicast congestion control algorithm enables reception rates at receivers to follow the familiar sawtooth pattern which arises when using TCP's AIMD congestion control. We facilitate this by providing two key contributions. First, we define a stair layer, a layer whose rate dynamically ramps up over time from a base rate of one packet per RTT up to a maximum rate before dropping back to the base rate. The primary benefit of this component is to facilitate additive increase automatically, without the need for IGMP control messages. Second, we provide an efficient hybrid approach to combine the benefits of cumulative and non-cumulative layering *below* the stair layer. This hybrid approach provides the flexibility of non-cumulative layering, while mitigating several of the performance drawbacks associated with pure non-cumulative layering. While our STAIR approach appears complex, 1) the algorithm is straightforward to implement and easy to tune, 2) it delivers data to each receiver at a rate that is in very close correspondence to the behavior of a unicast TCP connection over the same path, and 3) it does so with a quantifiable and reasonable bandwidth cost.

2   Definitions and Building Blocks for Our Approach

In order to motivate our new contributions, we begin with techniques from previous work which relate closely to our approach. In [3], four metrics for evaluating a layered multicast congestion control scheme are provided, two of which we recapitulate here.

Definition 1. *The* join complexity *of an operation (such as additive increase) under a given layering scheme is the number of multicast join messages a receiver must issue in order to perform that operation in the worst case.*

Definition 2. *For a layering scheme which supports reception rates in the range $[1, R]$, and for a given link l in a multicast tree, let $M_l \leq R$ be the maximum reception rate of the set of receivers downstream of l and let C_l be the bandwidth demanded in aggregate by receivers downstream of l. The* dilation *of link l is then defined to be C_l/M_l. Similarly, the dilation imposed by a multicast session on tree T is taken to be $max_{l \in T}(C_l/M_l)$.*

Table 1 compares the performance of various layering schemes which attempt to perform AIMD congestion control. Briefly, one cannot perform additive increase in a standard cumulative protocol[1], and while non-cumulative schemes can do so, they do so only with substantial control traffic and/or bandwidth dilation per operation.

Table 1. Performance of AIMD Congestion Control for Various Approaches.

Sequence	Dilation	Complexity of AI	Complexity of MD
Ideal	1	zero	zero
Std. Cum	1	N/A	1 leave
Std. NonCum	2	$O(\log R)$	$O(\log R)$
FGLM [3]	1.6	2 joins, 1 leave	1 leave

---

[1] Standard refers to the doubling scheme described in the introduction.

We briefly sketch one non-cumulative layering scheme used [3]. The layering scheme is defined by $B_0 = 1, B_1 = 2$, and $B_i = B_{i-1} + B_{i-2} + 1$ for $i \geq 2$. The first few rates of the layers for this scheme are 1, 2, 4, 7, 12, 20, 33,..., where the base rate can be normalized arbitrarily. Increasing the reception rate by one unit can be achieved by the following procedure: Choose the smallest layer $i \geq 0$ to which the receiver is not currently subscribed; then subscribe to layer i and unsubscribe from layers $i - 1$ and $i - 2$. A receiver can *approximately* halve its reception rate by unsubscribing from its highest subscription layer. While this does not exactly halve the rate, the decrease is bounded by a factor which lies in the interval from approximately 0.4 to 0.6.

One salient issue with FGLM is that the base layer bandwidth B_0 is fixed once for all receivers. Setting B_0 to a small value mandates frequent subscription changes (via IGMP control messages) for receivers with small RTTs. Setting it to be large causes the problems of abrupt rate increases and buffer overruns that FGLM is designed to avoid.

3   Components of Our Approach

In this section, we describe our two main technical contributions. The first contribution is a method for minimizing the performance penalty associated with non-cumulative layering by employing a *hybrid* strategy which involves both cumulative and non-cumulative layers. Our approach retains all of the benefits of fine-grained multicast advocated in [3], with the added benefit that the dilation can be reduced from 1.62 down to $1 + \epsilon$ with only a small increase in the number of multicast groups. The second contribution introduces new, dynamic *stair* layers, which facilitate fine-grained additive increase without requiring a substantial number of IGMP control messages. Taken together, these features make the fine-grained layered multicast approach much more practical.

3.1   Combining Cumulative and Non-cumulative Layering

In a conventional cumulative organization of multicast layers, only cumulative layers are used to achieve rates in the normalized range $[1, R]$.

- *Cumulative Layers (CL)*: The base layer rate is $c_0 = 1$, and for all other layers C_i, $1 \leq i \leq \log_\alpha R$, the rate $c_i = c_0 * \alpha^{i-1}$. When $\alpha = 2$, this corresponds to doubling of rates as each layer is added.[2].

In the fine-grained multicast scheme of [3], only non-cumulative layers are used to achieve a spectrum of rates over the same normalized range.

- *Non-Cumulative Layers (NCL)* : The non-cumulative layering scheme Fib1 presented in [3] has layers N_i whose rates are specified by the Fibonacci-like sequence $n_0 = 1, n_1 = 2$, and $n_i = n_{i-1} + n_{i-2} + 1$ for $i \geq 2$.

Note that both CLs and NCLs are *static* layers for which the transmission rate to the layer fixed for the duration of the session.

---

[2] Bandwidths can be scaled up multiplicatively by a base layer bandwidth B_0 in these schemes.

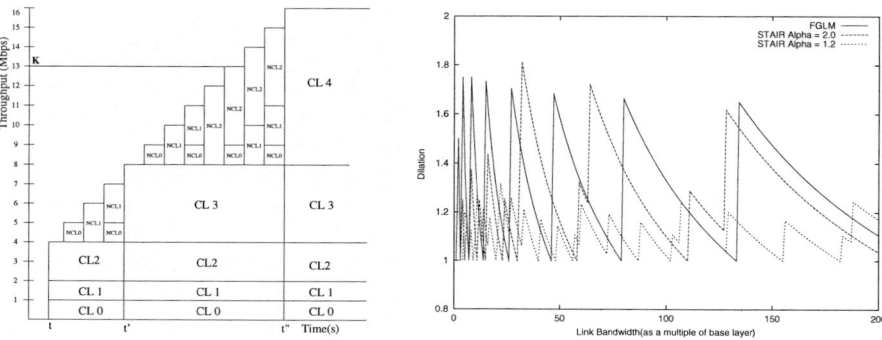

Fig. 1. (Left) Hybrid Layer Scheme : $K = \alpha^j + r$, $K = 2^3 + 5$ when $\alpha = 2$. CL denotes Cumulative Layer, NCL denotes Non-Cumulative Layer,
(Right) Maximal dilation at a link as a function of available link bandwidth.

In the hybrid scheme which we propose, we will require that *both* a set of cumulative layers C_i and a set of non-cumulative layers N_i are available for subscription. To attain a given subscription rate K, a receiver will subscribe to set of cumulative layers to attain a rate that is the next lowest power of α, capped by a set of non-cumulative rates to achieve a rate of exactly K, as depicted in Figure 1(left). In particular, we let $j = \lfloor \log_\alpha K \rfloor$ and write $K = \alpha^j + r$, then subscribe to layers $C_0, \ldots, C_j$ as well as the set of non-cumulative layers $\{N_r\}$ that the FGLM scheme would employ to attain a rate of r. As prescribed by FGLM, fine-grained increase (adding c_0) requires one join and two leaves, except for the relatively infrequent case when we move to a rate that is an exact power of α. In this case, we unsubscribe from all non-cumulative layers and subscribe to one additional cumulative layer. Multiplicative decrease now requires one leave from a cumulative layer and one leave from a non-cumulative layer. Comparing against a standard non-cumulative scheme, which used $\log_{1.6} R$ layers, we have now added $\log_\alpha R$ cumulative layers, or a constant factor increase. What we have gained is a dramatic improvement in dilation, expressed as the following lemma.

Lemma 1. *The dilation of the hybrid scheme is* $1 + 1.62\frac{(\alpha-1)}{\alpha}$.

Proof. We proceed by proving an upper bound on the dilation of an arbitrary link ℓ, which gives a corresponding bound on the dilation of the session. For each user U_j downstream of ℓ, consider the rate it obtains over cumulative layers a_j and the rate it obtains over non-cumulative layers b_j separately and denote its total rate by u_j. Let the user with maximal total rate $a_j + b_j$ be denoted by $\hat{U}$ and its rates be denoted by $\hat{a}$ and $\hat{b}$ respectively. Now reconsider user U_j. If $a_j < \hat{a}$, then by the layering scheme employed, $u_j = a_j + b_j < \alpha a_j$. Adding αb_j to both sides gives :

$$u_j + \alpha b_j < \alpha a_j + \alpha b_j = \alpha(u_j) \tag{1}$$

Simplifying yields: $b_j < \dfrac{u_j(\alpha-1)}{\alpha} \le \dfrac{\hat{u}(\alpha-1)}{\alpha}$

Otherwise, if $a_j = \hat{a}$, then by maximality $b_j \leq \hat{b} < (\alpha - 1)\hat{a}$. In either case, b_j is less than $\frac{\hat{u}(\alpha-1)}{\alpha}$, so $\max_j b_j$ is as well. From the dilation lemma proved in [3], a set of users subscribing to non-cumulative layers experience limiting worst-case dilation of 1.62. Thus the total bandwidth consumed by non-cumulative layers across ℓ is at most $1.62 \max_j b_j$. Plugging these derived quantities into the formula in Definition 2 yields:

$$\text{Dilation} \leq \frac{\hat{a} + 1.62 \max_j b_j}{\hat{a} + \hat{b}} < \frac{\hat{a} + 1.62 \frac{(\alpha-1)\hat{u}}{\alpha}}{\hat{a} + \hat{b}} \leq 1 + 1.62 \frac{(\alpha - 1)}{\alpha}. \qquad \square$$

Applying this lemma to a hybrid scheme with a geometric increase rate of $\alpha = 1.2$ on the cumulative layers realizes the benefits of a non-cumulative scheme, reduces the worst-case dilation in the limit from 1.62 to 1.27 (a 22% bandwidth savings) and requires only a modest increase in the number of groups. Figure 1 (right) shows the maximal dilation at a link as the link bandwidth varies as a function of n_0 for FGLM and the hybrid scheme for two different values of α. Recall that in FGLM, there are bandwidth *transition points*, when clients will subscribe to a new maximum layer j and unsubscribe from layers $j-1$ and $j-2$ across the bottleneck. At these transition points (spikes in the plot), worst-case dilation can be large due to the bandwidth consumed by this new layer. While STAIR with $\alpha = 2$ has comparable dilation to FGLM, STAIR with $\alpha = 1.2$ has substantially smaller worst-case dilation.

3.2   Introducing Stair Layers

Our next contribution is stair layers, so named because the rates on these layers change dynamically over time, and in so doing resemble a staircase. This third layer that a sender maintains is used to automatically emulate the additive-increase portion of AIMD congestion control, without the need for IGMP control traffic. Different stair layers are used to accommodate additive increase for receivers with heterogeneous RTT's from the source. These layers also "smooth" discontinuities between subscription levels of the underlying CLs and NCLs, which provide rather coarse granularity (in the subsequent discussion, we assume that these underlying layers have base rates $c_0 = n_0 = 1$Mbps). Finally, we note that the addition of stair layers increases the dilation beyond that proven in Lemma 1, but only by a small additive term, which we quantify in the full version of the paper.

– *Stair Layers (SL)*: Every SL has two parameters: 1) a round-trip time t in ms that it is designed to emulate and 2) a maximum rate R, measured in packets per t ms. The rate transmitted on each SL is a cyclic step function with a minimum bandwidth of 1 packet per t ms, a maximum of R, a step size of one packet, and a stepping rate of one per emulated RTT. Upon reaching the maximum attainable rate, the SL recycles to a rate of one packet per RTT.

Unlike CLs and NCLs, SLs are *dynamic* layers whose rates change over time. Dynamic layers were first used by [11] to probe for available bandwidth and

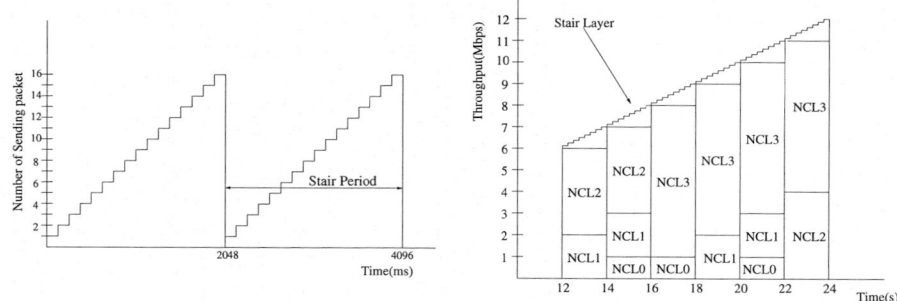

Fig. 2. Use of a Stair Layer with $t = 128$ms, $R = 1$Mbps, packet size $S = 1$KB. (Left) Rate of SL_{128} in isolation. (Right) SL_{128} used in conjunction with underlying non-cumulative layers.

later defined as such and used in [1] to avoid large IGMP leave latencies. Figure 2(left) shows the transmission pattern of SL_{128} (a stair layer for 128ms RTT) with maximum rate $R = 16$ packets per RTT. Also depicted in Figure 2 is a third useful parameter of a stair layer:

Definition 3. *The* stair period *of a given stair layer is the duration of time that it takes the layer to iterate through one full cycle of rates.*

Given a stair layer with an emulated RTT t and a maximum rate R the stair period p satisfies $p = Rt^2$. Typically, we will set the maximum rate R of a stair layer to be the base rate of the standard cumulative scheme c_0 (in Mbps), in which case we substitute for R and perform the appropriate conversions, assuming a fixed packet size S in bytes: $p = \left(\frac{c_0}{8S}\right) t^2$.

In practice, the sender will maintain several SLs to emulate a range of different RTTs. However, the fixed packet size and the maximum rate R of a stair layer give a lower bound on the range of RTT's that can be accommodated. The height of the staircase in steps directly corresponds to the factor in control traffic savings that will be achieved. Denoting this minimum desired height by h, we require that: $t \geq \frac{S*8*h}{R}$. For example, with a packet size of 512 bytes, $R = 1$ Mbps, and a desired value of $h = 8$, then the smallest allowable RTT in an SL is 32ms.

4   The STAIR Congestion Control Algorithm

We now describe how the techniques we have described come together into a unified multirate congestion control algorithm. We employ a hybrid scheme as described in Section 3.1, from which each receiver selects an appropriate subset of layers, used in concert with *one* stair layer, appropriate for its RTT. The two most significant challenges to address are providing the algorithms to performing additive increase and multiplicative decrease, respectively. Two additional challenges we address are 1) incorporating methods for estimation of multicast RTTs and 2) establishing a set of appropriate stair layers.

4.1 Additive Increase, Multiplicative Decrease

In order for a set of stair layers to complement a set of CLs and NCLs, the maximum rate of the stair layer must be calibrated to the base rate of the CLs and NCLs. The effect of appropriately calibrated rates can be seen in Figure 2: at exactly those instants when the stair layer recycles, the subscription rate on the NCL's increases by n_0, to compensate for the identical decrease on the stair layer. Now in order to conduct AIMD congestion control, the receiver measures packet loss over each stair period (during which additive increase takes place automatically). If there is no loss, then the receiver performs an increase of n_0, the base bandwidth of the NCLs as described earlier (1 join and 2 leaves or k leaves when the stair period is an exact power of α). As an aside, we note that it may be much more efficient for a last-hop router to handle such a *batch* of IGMP leave requests, rather than handling them as k separate requests.

Conversely, if there is a packet loss event in a stair period (of one or more losses), then one round of multiplicative decrease is performed. Approximately decreasing the rate by half is straightforward – it is necessary to drop the top cumulative layer as well as the top non-cumulative layer. While existing non-cumulative layering schemes do not easily admit dropping rates by exactly a factor of two, the consequences are mitigated substantially in a hybrid scheme; moreover, our experimental results indicate that the level of TCP-friendliness which can be achieved using our approach remains very high. We also note that there is no particular reason to wait until a stair period terminates before conducting multiplicative decrease – it can be done any time. Since the STAIR receiver unsubscribes and subscribes frequently to increase the rate, IGMP leave latency could be problematic. One solution is to perform joins and leaves in advance of when those operations need to take effect; we are also hopeful that subsequent versions of IGMP will accommodate fast IGMP leaves so that we can use them directly to respond to congestion in a timely fashion.

4.2 Configuration of Stair Layers

As motivated earlier, to accommodate a wide variety of receivers, stair layers must be configured carefully. We choose to space the RTTs across the available stair layers exponentially. Let RTT in the base Stair Layer be $2^i ms$. The base Stair layer increases its sending rate every $2^i ms$ and all the other stair layers j will increase the sending rate in every 2^{j+i} ms. The TCP throughput rate R, in units of packets per second, can be approximated by the formula in [10]: $R = \frac{1}{RTT\sqrt{q}(\sqrt{2/3}+6\sqrt{3/2}q(1+32q^2))}$, where R is a function of the packet loss rate q, the TCP round trip time RTT, and the round trip time out value RTO, where $RTO = 4RTT$ according to [5].

Since the throughput is inversely proportional to RTT, the receiver with a small RTT is more sensitive to the throughput than the receiver with large RTT, thus we recommend that RTTs provided by stair layers be exponentially spaced. Note that with an exponential spacing of stair layers, a receiver may subscribe to a different SL if its measured RTT changes significantly: it can can subscribe to a faster layer at the end of its current stair period, or drop down to a slower stair layer every other stair period.

4.3   RTT Estimation and STAIR Subscription

Each receiver must measure or estimate its RTT to subscribe to an appropriate stair layer. A variety of methods can be employed to do so; we describe two such possibilities, with the expectation that any scalable method can be employed in parallel with our approach. Golestani et al. provide an effective mechanism to measure RTT in multicast using a hierarchical approach [6]. However, their approach requires clock synchronization among the sender and receivers and depends on some router support which is not widely available. Another simple way to estimate RTT is to use one of various `ping`-like utilities. However, one cost associated with use of ping is that as the number of receivers increase the sender faces a "ping implosion" problem. We leave efficient RTT estimation for future work, noting the need for careful study of the tradeoff between frequency of measurement and accuracy of estimation.

Assuming that the receiver has an estimate of its RTT, its next challenge is to subscribe to the appropriate stair layers. Let RTT_i be the RTT in SL_i and RTT_m be the measured RTT. The receiver can subscribe to appropriate stair layers based on the measured RTT in the following way. If the RTT_m is within a $(1+\epsilon)$ factor from RTT_i for some i, simply subscribe to SL_i. A reasonable choice for ϵ which we argue for in the full paper is $\epsilon = 1/3$. Otherwise, to decrease the error bound in certain cases, a receiver should subscribe to the *two* smaller stair layers SL_i and SL_{i-1} for which $RTT_i < RTT_m$.

5   Experimental Evaluation

We have tested the behavior of STAIR extensively using the NS simulator [9]. The simulation results show that STAIR exhibits good inter-path fairness when competing with TCP traffic in a wide variety of scenarios. Our initial topology is the "dumbbell", with all non-bottleneck links set to 10ms delay and 100 Mbps bandwidth. In this topology, we vary the cross-traffic multiplexing level by varying the number of TCP flows, vary the bottleneck bandwidth, and scale the queue size. We then consider the impact of richer topologies, including multiple bottleneck links, and TCP cross traffic with both short and long RTTs. Throughout our experiments, we set $c_0 = n_0 = 512$ Kbps and set $\alpha = 2$, i.e. the rate $c_i = 2^{i-1} * 512$ Kbps for $i > 0$. We employ a fixed packet size of 512B throughout. Also, while there is theoretical justification for smaller settings of α, we did not observe worst-case dilation often in our simulations.

In most the experiments we describe here, we use RED gateways, primarily as a source of randomness to remove simulation artifacts such as phase effects that may not be present in the real world. Use of RED vs. drop-tail gateways does not appear to materially affect performance of our protocol. The RED gateways are set up in the following way: we set the queue size to twice the bandwidth-delay product of the link, set *minthresh* to 5% of the queue size and *maxthresh* to 50% of the queue size with the *gentle* setting turned on. Our TCP connections use the standard TCP Reno implementation provided with NS.

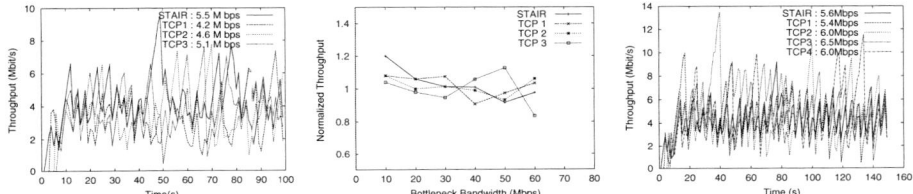

Fig. 3. (a)(b): TCP Flows and One STAIR with RED, (c) DropTail.

Figure 3 shows the throughput of a one receiver STAIR flow competing with three TCP Reno flows on RED. Figure 3(a) shows the throughput of STAIR competing with three TCP flows across a dumbbell with a 12ms/20Mbps bottleneck link. In this environment, STAIR fairly shares the link with TCP flows. We next vary the bandwidth of the bottleneck link to assess the impact of changing the range of subscribed NCLs. Figure 3(b) shows the average throughput trends achieved by three long-running TCP flows and one STAIR flow on various bottleneck bandwidths.

Figure 3(c) shows the throughput of STAIR competing with four TCP Reno flows on drop-tail gateways competing for 30Mbps bottleneck bandwidth. Dynamics across a bottleneck drop-tail gateway tended to be more dramatic, but overall fairness remained high. Here, the throughputs of TCP receivers ranged between [5.4Mbps, 6.5Mbps] with a mean per-connection throughput of 6.0 Mbps, while the STAIR receiver had average throughput of 5.6Mbps.

We used a second topology to test heterogeneous fairness (see Figure 4) under different RTTs. We consider a single STAIR session with two STAIR receivers and two parallel TCP flows sharing the same bottleneck link. The RTT of STAIR receiver 1, R_{s1}, is 60ms, while the RTT of STAIR receiver 2, R_{s2}, is 120ms. In our experimental set up, each receiver periodically samples the RTT using `ping`. R_{s1} subscribes to SL_{64}, which is the closest stair layer based on the measured RTT, while R_{s2} subscribes to SL_{128}. The throughput of each of the flows is plotted in Figure 4. Both of the STAIR flows share fairly with the parallel TCP flows with the same RTT. Since the throughput of TCP is inversely proportional to RTT, the receiver R_{s2} should have approximately half of R_{s1}'s average throughput. In

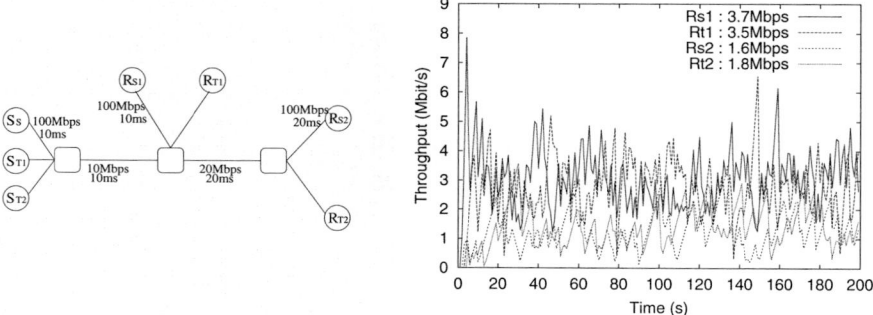

Fig. 4. Throughput of STAIR and TCP flows sharing bottleneck link with different RTT,R_{S1}, R_{S2} : STAIR Receiver, R_{T1}, R_{T2} : TCP Receiver.

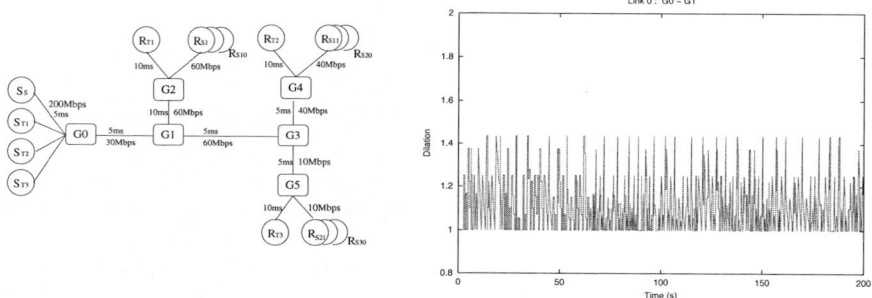

Fig. 5. Dilation on the Bottleneck link(G0 - G1).

this experiment, the average throughput attained by R_{s1} was 3.7Mbps and the average throughput attained by R_{s2} was 1.6Mbps.

We then increased the number of STAIR receivers using the topology in Figure 5. The starting time of 30 STAIR receivers are uniformly distributed from 1 second to 10 second. The average throughput attained were : $R_{s1} \dots R_{s10}$: 8.6 Mbps , R_{t1}: 7.3 Mbps, $R_{s11} \dots R_{s20}$: 8.7 Mbps, R_{t2}: 8.27Mbps, $R_{s21} \dots R_{s30}$: 4.5Mbps, R_{t3}: 2.3 Mbps. The discrepancy between receiver R_{s30} and R_{t3} points to an aspect of TCP behavior we have not yet captured accurately. When a STAIR receiver subscribes a new maximum j and unsubscribes from layers $j-1$ and $j-2$, it can cause a significant increase in bandwidth consumption. Although the measured dilation on the link(G3 - G5) is less than the dilation in Figure 5, the increases were substantial enough to drive TCP flows into timeout. Since TCP timeout behavior is not yet accurately reflected in STAIR, some unfairness can result.

We then considered varying the queue size of the bottleneck link router in our baseline topology, holding all other parameters constant. To make this simulation more interesting, we used drop-tail gateways to magnify the negative

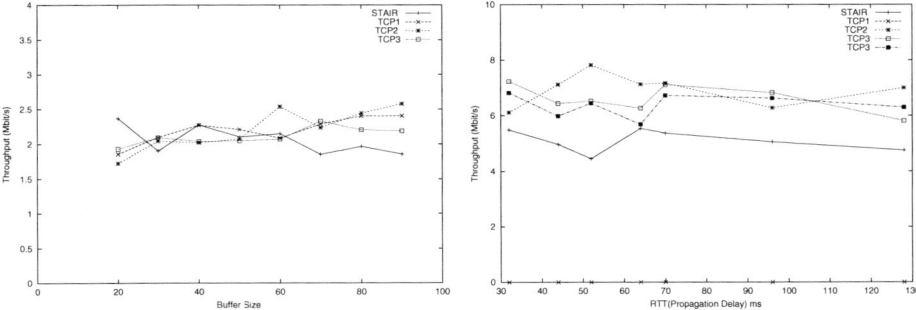

Fig. 6. Throughput on Different Queue Size (left) and Different RTT (right), with DropTail.

performance impact of large queues. Note that for these simulations, when the queue size is large (overprovisioned with respect to the bandwidth-delay product of the link), the RTT is affected by queuing delay. STAIR receivers adapt by changing the stair layer depending on the measured RTT. Figure 6 shows the throughput of STAIR and TCP as we vary the queue size. When the RTT varies over time, the throughput is affected by the error bound of RTT. Even though the average throughput is reduced as the queue size increases, STAIR is not especially sensitive to queue sizes, unlike some other schemes. Finally, we consider varying the link delay on the bottleneck link. Figure 6 (right) shows that as the estimated RTT increases, STAIR becomes less aggressive in accordance with TCP. Additional experiments which we conducted are available in the full version of this paper [2].

6   Conclusions

We have presented STAIR: a hybrid of cumulative, non-cumulative and stair layers to facilitate receiver-driven multicast AIMD congestion control. Our approach has the appealing scalability advantage that it allows receivers to operate asynchronously with no need for coordination; moreover, receivers with widely differing RTTs may simulate different TCP-friendly rates of additive increase. While asynchronous joining and leaving of groups at first appears to run the risk of consuming excessive bandwidth through a shared bottleneck, in fact judicious layering can limit the harmful impact of this issue.

Our approach does have several limitations, which we plan to address in future work. First, while our congestion control scheme tolerates heterogeneous audiences, it is primarily designed for users with high end-to-end bandwidth rates in the hundreds of Kbps range or higher. We expect that slower users would wish to employ a different congestion control strategy than the one we advocate here. Second, congestion control approaches which use non-cumulative layering and dynamic layers cannot be considered general purpose (just as TCP's congestion control mechanism is not general-purpose) since not all applications can take full advantage of highly layer-adaptive congestion control techniques. For now, the only application which integrates cleanly with our congestion control methods is reliable multicast of encoded content. We hope to develop scalable STAIR methods compatible with other applications, such as real-time streaming.

References

1. J. Byers, M. Frumin, G. Horn, M. Luby, M. Mitzenmacher, A. Roetter, and W. Shaver. FLID-DL: Congestion Control for Layered Multicast. In *Proceedings of NGC 2000*, pages 71–81, November 2000.
2. J. Byers and G. Kwon. STAIR: Practical AIMD Multirate Multicast Congestion Control, Technical Report BUCS-TR-2001-018, Boston University, Sept. 2001.
3. J. Byers, M. Luby, and M. Mitzenmacher. Fine-Grained Layered Multicast. In *Proc. of IEEE INFOCOM 2001*, April 2001.
4. J. Byers, M. Luby, M. Mitzenmacher, and A. Rege. A Digital Fountain Approach to Reliable Distribution of Bulk Data. In *Proc. of ACM SIGCOMM*, pages 56–67, 1998.
5. S. Floyd, M. Handley, J. Padhye, and J. Widmer. Equation-based congestion control for unicast application . In *Proc. of ACM SIGCOMM*, 2000.
6. S. Golestani. Fundamental observations on multicast congestion control in the Internet. In *Proc. of IEEE INFOCOM'99*, New York, NY, March 1999.
7. A. Legout and E. Biersack. PLM: Fast convergence for cumulative layered multicast transmission schemes. In *Proc. of ACM SIGMETRICS 2000*, pages 13–22, Santa Clara, CA, 2000.
8. S. McCanne, V. Jacobson, and M. Vetterli. Receiver-Driven Layered Multicast. In *Proc. of ACM SIGCOMM'96*, pages 1–14, August 1996.
9. ns: UCB/LBNL/VINT Network Simulator (version 2). Available at http://www-mash.cs.berkeley.edu/ns/ns.html.
10. J. Padhye, V. Firoiu, D. Towsley, and J. Kurose. Modeling TCP throughput: A simple model and its empirical validation. In *Proc. of ACM SIGCOMM*, 1998.
11. L. Vicisano, L. Rizzo, and J. Crowcroft. TCP-like Congestion Control for Layered Multicast Data Transfer. In *Proc. of IEEE INFOCOM'98*, April 1998.
12. J. Widmer, R. Denda, and M. Mauve. A Survey on TCP-Friendly Congestion Control. *IEEE Network*, May 2001.
13. K. Yano and S. McCanne. The Breadcrumb Forwarding Service: A Synthesis of PGM and EXPRESS to Improve and Simplify Global IP Multicast. In *Proc. of ACM SIGCOMM Computer Communication Review (CCR), 30 (2), April*, 2000.

Impact of Tree Structure on Retransmission Efficiency for TRACK

Anthony Busson*, Jean Louis Rougier, and Daniel Kofman

Ecole Nationale Supérieure des Télécommunications
46 rue Barrault, 75013 Paris, France
{abusson,rougier,kofman}@enst.fr

Abstract. This paper focuses on tree based reliable multicast protocols, or more precisely TRACK (Tree based Acknowledgment) protocols, as defined by IETF. With the TRACK approach, the classical feedback implosion problem is handled by a set of servers, organized in a tree structure, which are in charge of local retransmissions and feedback aggregation. We study the impact of the control tree structure (for instance the number of servers to be deployed) on transmission performances.
We propose a new model, where point processes represent the receivers and the servers, which captures loss correlation phenomena. We are able to get explicit expressions of the number of useless retransmissions (useless as the given segment was already received) as a function of a limited number of tree characteristics. Generic tree configuration rules, optimizing transmission efficiency, are obtained.

1 Introduction

1.1 Reliable Multicast Issues

IP multicast leads to excellent network utilization when dealing with groups communication, as packets sent by the participants are duplicated by the routers for the different receivers. Important advances have been made in the multicast routing protocols. Multicast applications using IP multicast are becoming widely available, and are beginning to be commercialized. These group communication applications need to transfer data with a certain degree of reliability. However, no multicast transport layer has been standardized as yet. Intense research is thus being undertaken to develop multicast transport protocols. The main problems with reliable multicast protocols are due to scalability requirements as multicast sessions may contain a very large number of receivers (news networks, sports events,etc.). The main factor affecting the scalability of multicast reliable transport protocols is due to the so-called "feedback implosion" problem: if all receivers acknowledge the received segments, the source would quickly become overflooded with such feedback information as the number of receivers grows. On the other hand, the source actually needs to know the receivers window state

---

* This work is supported by the french government funded project GEORGES (http://www.telecom.gouv.fr/rnrt/projets/pgeorges.htm).

J. Crowcroft and M. Hofmann (Eds.): NGC 2001, LNCS 2233, pp. 113–127, 2001.
© Springer-Verlag Berlin Heidelberg 2001

in order to retransmit lost segments, and some receiver statistics (RTT, explicit or implicit congestion indications,...) for congestion control. Several approaches have been designed, and to some extent tested in the Mbone, in order to solve this feedback implosion.

- Negative ACKnowledgements can be used (the so-called NORM : Nack Oriented Reliable Multicast inspired being [1]). Such approach make use of timers in order to avoid many NACKs being sent for the same data segment : when a packet is detected missing, the receiver triggers a backoff timeout period, after expiration of the timeout, if the receiver has not received a repair for the lost data or another NACK from other receivers (in the case where NACK are multicast) the receiver sends its NACK to the source. Such "flat" approach is very interesting for its simplicity, minimal configuration and coordination requirements between members of a multicast session, but suffers from scalability issues.
- Network nodes can be used in order to improve the scalability and efficiency of reliable multicast protocols. For instance, in Pragmatic General Multicast (PGM, [2]), a receiver which detects a missing packet multicasts a NACK. A router which implements the PGM protocol (called a PGM router) immediately multicasts a suppression message on the same port in order to avoid multiple NACKs from receivers. The operation is repeated until the source is reached, the recovery packet is then multicasted only on ports which have received a NACK. Generic Router Assistance (GRA, [3]) is being studied and standardized at IETF, however such mechanisms may be difficult to deploy, as existing routers do not support such services.
- Tree based protocols (TRACK : TRee-based ACKnowledgement [4],[5]) are extensively scalable by dividing the set of receivers into controlled subgroups, a dedicated server being responsible for local retransmissions and feedback aggregation for its subgroup. Further scalability can be obtained by organising servers in subgroups and so on, generating a control tree with the source (or a specific dedicated server) as root. TRACK architectures require more configuration as compared to the previous approaches (in order to set the control tree), but it appears to be the most extensively scalable solution for reliable multicast (as for instance some experiments on RMTP-II protocols [6], and analytical sudies [7] [8] have shown that trees are an answer to the scalability problems). In this paper, we shall concentrate on dimensioning rules for tree configuration of such TRACK approaches.
- FEC (Forword Error Control) can be used in order to limit receiver acknowledgement (or even suppress the need of feedback for "semi-reliable" servivces). Note that FEC may be used in conjunction with the previous approaches, particulary with TRACK and NORM. We shall not consider this case in the present paper.

Another big issue regarding the development of reliable multicast protocols is the design of congestion control algorithms. For instance, a main difficulty here is the heterogeneity of receiver access bit rates and processing capacity. We shall concentrate on the feedback implosion issue, congestion control is out of scope of this paper.

1.2   Contribution

In this paper, we study the impact of the control tree structure on the retransmission efficiency. In [9], we proposed a geometrical model using point processes in the plane organized in clusters (representing the receivers location concentration such as in AS[1], POPs[2], etc.). The point processes allowed us to introduce a heterogeneity of the loss distribution among the set of receivers and thus to take loss correlation into account. The model presented had strong constraints as we distributed servers (used in TRACK) within each concentration domain. The choice of new point processes which describe the servers allow us in this paper to removed these constraints. We define a new cost function representing the number of packets received by the participants when they have already received it (which we shall call "useless retransmisions"). We believe that this new cost function is more realistic with regards to the traditional number of retransmissions (used in [9–11]). Stochastic geometry [12, 13] has already been proved to be adapted for the study of multicast routing protocols [14].

We find explicit formulae for the average cost for a generic distribution of receivers within a cluster, and compute the optimal parameters of the tree in terms of the number of children per parent. Finally, dimensioning rules are given with regard to the parameters describing the macroscopic behavior of the session, such as the loss probability inside and outside the concentration domains, the total number of receivers and the number of domains.

The paper is organized as follows: in Section 2, we briefly describe the TRACK protocols. The mathematical model, and the cost function are explained in Section 3 and 4. Numerical results are given in 5. The conclusion is presented in Section 6. For the sake of clarity, we give the computation details and explicit formulae in the Appendix.

2   TRACK

The main issue regarding the development of multicast transport protocols is related to scalability. Indeed, for an ARQ based (Automatic Repeat reQuest) protocol, each receiver send its own reception state, either periodically by ACKs[3] or only when a packet is detected missing (NACK). In both cases, the source can be flooded by this feedback information. The TRee-based ACK architecture (TRACK) [4] [5] avoids this problem by constructing a hierarchy of servers which aggregate feedback information of the receivers. Receivers do not send their state to the source but to a server called a parent, which aggregates this information and repairs the eventual lost packets locally.

---

[1] Automous System.

[2] Point of Presence.

[3] A bit map of received packets is periodically sent to the source.

TRACK is designed to reliably send data from a single sender to a group of receivers. At the beginning of the multicast session, each receiver bind to a near Repair Head (RH). This Repair Head is defined as being responsible for local error recovery and receiver state aggregation. Possible mechanisms that allow a receiver to know the list of available RHs, and to build the resulting tree are given in [15] and [16]. Repair Heads of different receivers (children) are bound to a RH of a superior level (parent). This binding is repeated until the source is reached. By convention, the source is level 0 (and is considered as the top RH), and the level of children is one more than the one of their parent.

2.1 Tree Structure

We shall distinguish two communication channels: the *data channel* used to multicast data from the source to the members of the group and the RHs, and the *control tree* used to aggregate information and repair lost packets. The source sends original data on the data channel only once. In the control tree, a RH (or the sender) uses a multicast address to perform local retransmissions and to send maintenance packets to its children. In order to limit the retransmission scope of a RH to the set of its children, each RH repair lost packet with a different multicast address. In figure 1(a), we show a data channel and a control tree with three levels.

2.2 Repair Head Functions

Once the control tree is built, it can be used for retransmissions and receiver feedback aggregation. The receiver feedback is necessary for determining eventual lost packets (for RH retransmissions) and statistic collection: these statistics (RTT, loss rates, number of children,etc.) are aggregated by each RH and passed upwards. The source can use these indications for its congestion algorithm (which is outside the scope of this paper). In TRACK, receiver feedback consists of regular ACKs, and optionnally NACKs. In both cases mechanisms are used to control and limit the amount of feedback received by a RH. For ACKs, all the receivers unicast their reception states at a different time, ensuring a constant (and controlled) load to the RH. In case of NACKs, timer based schemes are used to avoid unnecessary messages. When a given segment is detected missing, a receiver triggers a random timer. The NACK is sent only if the timer expires and if no retransmission for the same segment is received during this time interval.

2.3 Study of TRACK Protocols

A variety of studies exist which analyse the behavior of a multicast session. A recent work ([17]) analyses the throughput of a one-to-many communication with regards to the topology of the routing tree and the number of receivers. In [10], the placement of hierarchical reliable multicast servers based on tree knowledge is optimized. We believe that in many cases, this won't be possible (dedicated servers have to be placed in advance). TRACK dedicated servers will most probably be deployed without knowledge about receiver location nor, therefore, the data multicast tree (as in the mesh approach in [16]).

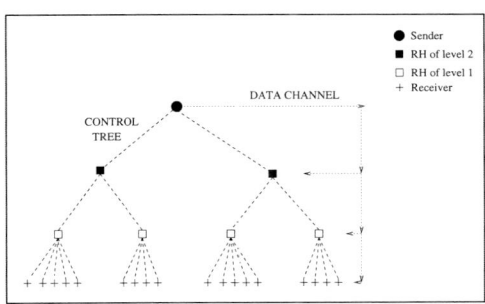 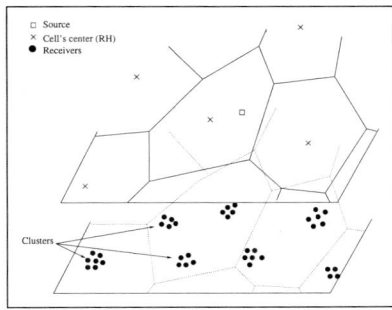

(a) Data channel and control tree.

(b) The model: Clusters in Voronoi cells.

Fig. 1. Control Tree and the Geometric Representation.

3   The Analytical Model

In this section, we present the mathematical framework used to model a TRACK session. Our approach is based on two models:

- A geometrical model described on Section 3.1 using point processes in the plane which represent the location of participants of a multicast session for a single source and the set of RHs,
- A loss model described on Section 3.2 which represents the distribution of losses in the network.

3.1   Geometrical Model

Receivers: In order to model the hierarchical structure of the Internet, where receivers are concentrated in local regions (e.g. site, AS,etc.), we have chosen the cluster point process described below to represent the set of receivers participating in the multicast session. In such a process, a first point process is scattered in the plane. Each point of this process generates a cluster of points, generating a new point process (points of the first process are usually no longer considered). In the case we shall study, the underlying point process is a homogeneous Poisson process (the choice of such a process will be discussed in Remark 1). It represents the set of locations of the concentration domains. The Clusters are i.i.d. (independent and identically distributed) point processes, and represent the receivers within a concentration domain. The resulting processes are called Neyman-Scott processes, and are shown to be stationary ([23]). Formally, we shall denote by π_{cl} the Poisson point process which generates the set of clusters and N_x the point process which represents the cluster located at x. With these notations, the so-defined Neyman-Scott process N defined above can be written:

$$N = \sum_{x \in \pi_{cl}} N_x.$$

We do not define here the distribution of the clusters, as the formulae (see the Appendix) are calculated for a generic distribution. In Figure 2(a), a sample of a Neyman-Scott process is shown.

Repair Heads: The set of RHs is modeled on a homogeneous Poisson point process π_{RH}. A RH is a parent of receivers of several clusters.

We shall further assume that π_{RH} is independent of π_{cl} and N. Actually, we assume that the set of RHs are placed in the Internet in advance, for a large set of different multicast sessions. The RH locations are thus independent from the location of receivers that may join or leave groups at any time. Computing average costs (w.r.t. receiver distribution and RH placement) will thus lead to average TRACK performance for a wide range of possible group attendance distributions.

For each RH, let $V_y(\pi_{RH})$ be the Voronoi Cell centered at $y \in \pi_{RH}$, defined as the set of points of $\mathbb{R}^2$ which are closer to y than to any other point of π_{RH}. The set of Cells of π_{RH} forms a tesselation of the plane (a poisson process and its tesselation is shown in Figure 2(b)). We assume that receivers connect to their closest RH (i.e. closest point of π_{RH}). Thus, the children of $y \in \pi_{RH}$ are the set of receivers within the clusters for which the generating point (a point of π_{cl}) is inside the cell $V_y(\pi_{RH})$. Formally, all the points of a cluster located at z are bound to a point y of π_{RH} if and only if $z \in V_y(\pi_{RH})$ (see Figure 1(b)).

We will look at the restrictions of the processes on a finite window in order to keep a finite number of receivers, clusters and RHs. This window will be a ball in $\mathbb{R}^2$ of radius R_w.

3.2   Loss Model

The losses in a multicast tree are often correlated. Indeed, if a loss occurs in a node of the multicast tree, all participants which are downstream will not receive this packet. The number of participants which do not receive a packet depends on the location of the loss in the multicast tree. Yajnik et al. [22] have studied the distribution and the correlation of packet losses in the Mbone. They propose a loss model called a modified star topology. In such a model, a packet is lost for all receivers (i.e. near the source) with probability p, and may also be lost on the way to each receiver (independently of each other) with probability q. The probability of reaching a receiver is then $(1-p)(1-q)$. We have extended this modified star topology in order to fit it to the network model presented in section 3.1.

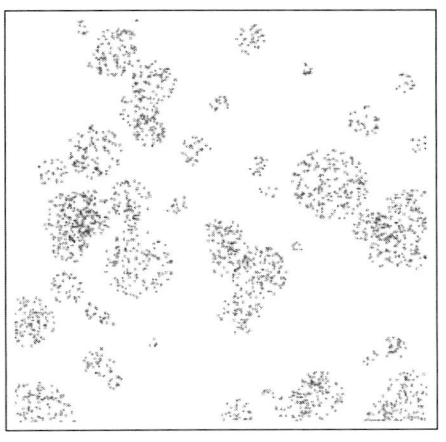

(a) Neyman-Scott process.

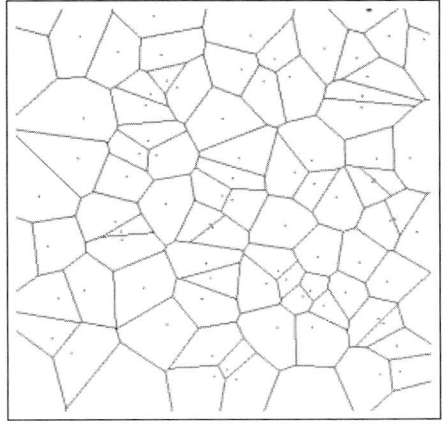

(b) Poisson point process and its tesselation.

Fig. 2. Point Process in the Plane.

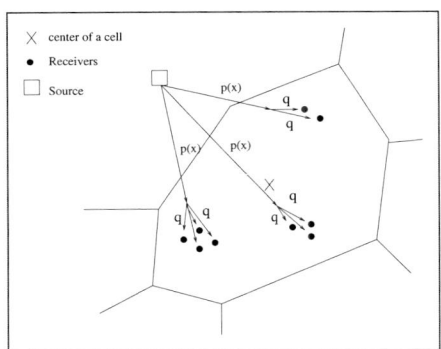

(a) loss model for a packet from the source.

(b) loss model for a retransmission from the center of a cell.

Fig. 3. The Hierarchical Loss Model.

Our loss model is the following:

– For the original transmission from the source
  • The probability of reaching any cluster (centered at $y \in \pi_{cl}$) is $1 - p(\|y\|)$, i.e. is a function of the distance betweeen the source and the cluster. For the sake of simplicity, we shall assume that $p(\|y\|) \simeq p(\|x\|)$ for all $y \in V_x(\pi_{cl})$, i.e. that the probability p depends on the distance between the source and the center of the cell to which the cluster belongs. However, we shall still assume that losses between the different clusters in a given cell are independent from each other.
  • If the packet has reached the cluster, the probability of reaching a receiver or a RH is $1 - q$ and is independent of other receivers.
– For a retransmission from the center of a cell (i.e. from a RH)
  • The retransmission from a RH $x \in \pi_{RH}$ reaches a cluster (i.e. domain, AS...) $y \in \pi_{cl}$ with probability $p(\|y - x\|)$ (which depend on the distance between the RH and the cluster). For the sake of simplicity, we shall approximate each loss probability $p(\|y - x\|)$ with $p(m)$ where m is the average distance between a cluster and the center of the closest cell (i.e. between the cluster and its RH). From [12], $m = \frac{1}{2\sqrt{\lambda_{RH}}}$ where λ_{RH} is the intensity of π_{RH} .
  • if the packet has reached the domain, the probability of reaching a receiver or a RH is $1 - q$ and is independent of other receivers.

In summary, for the original transmission from the source, a packet reaches a receiver in cell x with probability $(1 - p(\|x\|))(1 - q)$. For a retransmission from a RH (centers of the cells), a packet reaches a receiver with probability $(1 - p(m))(1 - q)$.

We could have added the probability that a packet has been lost for all participants of the multicast session. This case would correspond to a loss occured near the source. However, the cost function introduced in Section 4 (which will be the number of useless transmissions) would not depend on this parameter. Indeed, we shall consider a retransmission is useless for a given receiver (resp. RH) whenever a receiver (resp. RHs) receives a packet which it has already received it.

Choice of $p(\|x\|)$: we can consider that the probability of reaching a point x from an other point y is a function of the number of crossed domains. Moreover, if we suppose that the number of crossed domains is proportional to the distance between these two points (mathematical evidences of this property hold for different models, e.g. [21]), then we can choose $1 - p(\|x\|)$ such that it represents the probabibilty of reaching x from the source when each crossed domain is successfully crossed with a given probability (that we shall denote $1 - a$). Thus, if we denote by $N(x - y)$ the number of crossed domains between x and y, we have:

$$1 - p(\|x - y\|) = (1 - a)^{N(x-y)}.$$

Table 1. Loss Probability.

		mean loss
	0.1	0.625
a	0.01	0.085
	0.001	0.0088
	0.0001	0.00088

where $N(x - y) = \gamma \|x - y\|$ $(\gamma > 0)$.

In the numerical applications presented in Secion 5, clusters are associated with AS and $N(x)$ is chosen to represent the mean number of crossed domains in the Internet (i.e. related to the so-called BGP AS path length, that we denote by β). As discussed below other representations are possible. The parameter γ is then chosen such that the number of crossed domains for a typical point of π_{cl} be equal to β $(\gamma \mathbb{E}[\|y\|] = \beta, y \in \pi_{cl})$. In the Table 3.2, we give the mean loss probability to reach a typical cluster for a packet sent by the source w.r.t. a.

Remark 1. *In the Mathematical model defined above, clusters are bound to the closest RH w.r.t. the euclidian distance. In fact, since in our model $p(d)$ is increasing with the distance d, the clusters are connected to the RH which minimizes the loss probability. Here the loss probability has been considered, but it would be possible to choose other metrics (as proposed in [16]) such as delay, number of hops,etc... Moreover, it should be noted that the cost function does not depend directly on the location of points (which is just a mathematical representation), but depends solely on the distribution of losses. The function $p(d)$ allows a mapping between actual average loss probabilities and a two dimensional representation of receivers based on clusters in the plane.*

When more sophisticated statistical inference of multicast losses will be available (see [18], for instance), an interesting research issue will be to try to find more realistic mappings $p(d)$ and network representations. Nevertheless, due to lack of statistics, the choice of the Poisson distribution for clusters and RHs is motivated by its simplicity.

4   The Cost Function

The cost of a reliable multicast is difficult to evaluate as different metrics can be optimized (throughput, number of retransmissions, etc.). In this paper, we do not use the classical number of retransmissions. Indeed, retransmissions of a packet are useful while all receivers have not received the packet. Typically, when a loss occurs, a great number of the totality of receivers may not have received the packet. We propose a new cost function based on the number of "useless" transmissions. If a RH retransmits a packet, the number of useless transmissions is the number of receivers which will receive this packet when they have already received it (directly from the source or in a previous retransmission). Since a

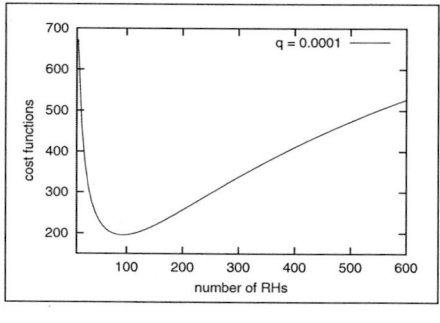

(a) A cost functions.

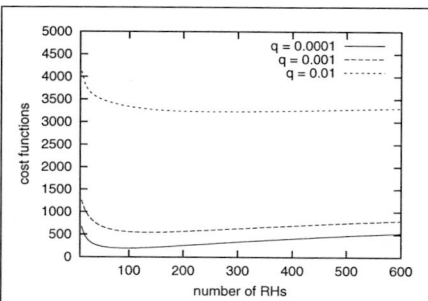

(b) The evolution of the cost function.

Fig. 4. Cost Functions.

retransmission from a RH is multicasted, there are always useless retransmissions unless all the children of this RH have not received this packet before.

The cost function is defined as the number of useless retransmissions for each level of hierarchy. Let Ξ_i be the random variable which represents the number of useless retransmissions in the level i, we have:

$$Cost = \sum_{i=1}^{2} \mathbb{E}[\Xi_i],$$

where $\mathbb{E}[\Xi_i]$ is the expectation of Ξ_i.

We note that we could weight each level of this function by constants in order to favour retranmission in a particular level rather than an other. For instance, since the servers have a cost, it can be interesting to favour useless transmission between the RHs and the receivers.

5   Results

For the sake of clarity, the computation details of the cost function and its explicit formulae are given in the Appendix. These formulae are given for a generic distribution of the number of receivers within a cluster. We consider a particular case of this distribution in Subsection 5.1 for which we compute the optimal value of the intensity of π_{RH} w.r.t. the loss parameters, the number of domains and the mean number of receivers per cluster.

5.1   Preliminaries

We do not need the exact point locations within a cluster, since the loss probability within a cluster is a constant q. We just have to choose the distribution

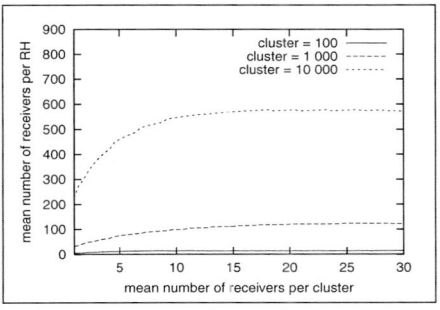
(a) The number of receivers per RH.

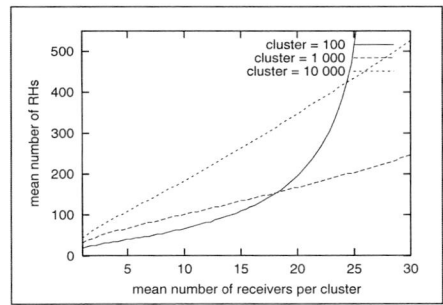
(b) The total number of RHs.

Fig. 5. Optimal Parameters when the Mean Number of Receivers Varies.

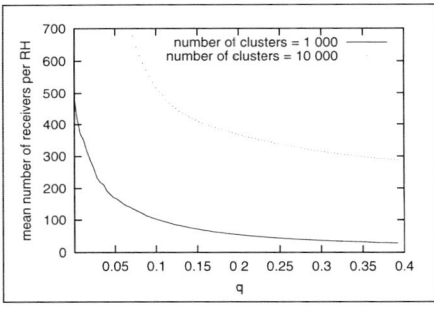
(a) The number of receivers per RH when q varies.

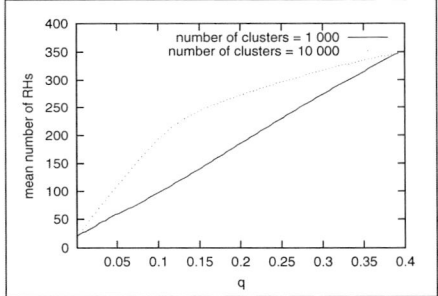
(b) The total number of RHs when q varies.

Fig. 6. Optimal Parameters when the Loss Probability q Varies.

of the number of receivers within a cluster. In this paper, the random variable which describes this number is Poisson distributed, chosen for its simplicity. In futur works, other distributions could be considered based on multicast session popularity for instance (e.g. [19, 20]).

5.2   Numerical Results

We give the optimal number of RHs when the number of receivers per cluster, the probability q and the parameter a (see Section 3.2) vary. In the following, when parameters are not specified, the default values are: 0.001 for the loss probability a and q and 10 for the mean number of receivers in a cluster.

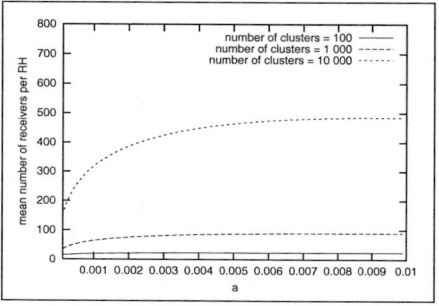
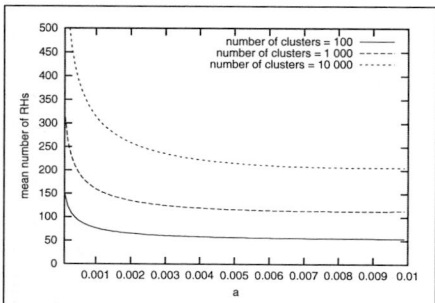

(a) The optimal number of receivers per RH.

(b) The total number of RHs.

Fig. 7. Optimal Parameters when the Loss Probability a Varies.

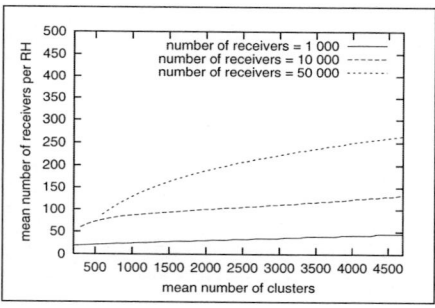
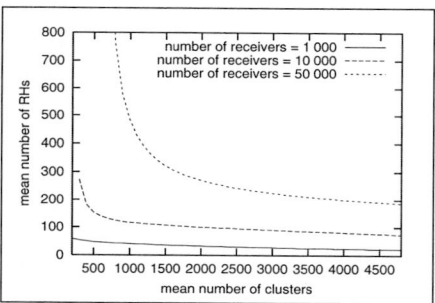

(a) The optimal number of RHs when the number of clusters varies.

(b) The total number of RHs when the number of clusters varies.

Fig. 8. Optimal Parameters when the Number of Clusters Varies.

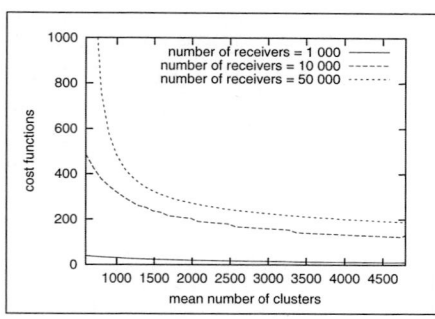

Fig. 9. The Cost Functions when the Number of Clusters Varies.

The optimal is found for the intensity of π_{RH} which minimizes the cost function. We deduce the optimal mean number of receivers per RH, the mean number of active RHs[4] and the total number of RHs (active or not).

In Figure 4(a) and 4(b), we show the cost function with regards to the total number of RHs. Figure 4(b) shows that the cost function can be very flat around its minimum, and the minimum can even be reached for an infinite number of RHs (since the function decreases). In this case, each cluster is connected to a different RH; the number of active RHs is then equal to the number of clusters.

Impact of the Number of Receivers. For a fixed number of active clusters (resp. 100, 1000 and 10000), Figure 5(b) represents the optimal number of RHs as a function of the number of receivers per cluster. Not surprisingly, the optimum number of RH increases as the number of receiver per cluster grows (the total number of receivers grows proportionally). It can be noted that for 100 clusters, the optimal number of RH diverges and tends to infinity. Of course, it does not make sense to choose an infinite number of RHs (only a finite number of RHs will be selected by the receivers in any case), this phenomenon corresponds to the fact that the optimum is reached when there is exactly one RH per domain. For a larger number of clusters however, it is interesting to note the optimal number of RHs is quasi-linear w.r.t the mean number of receivers per cluster, which will facilitate the determination of dimensioning rules.

In Figure 5(a), the optimal number of children connected to a RH is plotted (under the same conditions as above). It can be remarked that the optimal number of receivers bound to a RH increases as the cluster size increases. The optimum quickly reaches a plateau (which is related to the linearity of curves in Figure 5(b)).

Impact of the Intradomain Loss Probability (Parameter q). As expected, when the loss probability within a cluster q increases, the required number of RHs also increases (Figures 6(a) and 6(b)). The case where the mean number of clusters is 100 has not be drawn because the optimal intensity is infinite; the number of required servers is always one per cluster. We remark that even in the case where the loss probability q is very high (until 0.4), a RH is still responsible of a large number of receivers.

Impact of the Interdomain Loss Probability (Parameter a). When the probability a increases, the cost between the source and its children grows. Therefore, the number of children bound to a RH increases (Figure 7(a)) and the total number of RHs decreases (Figure 7(b)). The number of children per RH quickly reaches a plateau, we note that these values are quite similar to the Figure 5(a).

Impact of Loss Correlation. In this paragraph, we concentrate on the impact of receiver concentration (and thus of loss correlation). For a fixed mean number of receivers, we vary the mean number of clusters, and we change the number

---

[4] RHs which have at least one children.

of receivers per cluster accordingly. It can be noted that the loss correlation decreases as the number of clusters increases (the more clusters, the less receivers per cluster). In Figure 9, the optimal cost function is plotted as a function of the mean number of clusters. As expected, loss correlation has an important impact on transmission efficiency. It can be observed that the cost decreases as the loss correlation decreases, which can be explained by the choice of the cost function: When losses are correlated (i.e. a small number of large clusters), a retransmission will be useless for a larger number of receivers (almost all the nodes of one or several clusters). In Figure 8(a), the mean number of receivers bound to a RH is plotted as a function of the number of clusters. We can observe that the less important the loss correlation is (i.e. the larger the number of clusters), the less RHs are required. In Figure 8(b), we can observe that the optimum number of receivers per RH increases with the number of clusters (i.e. as loss correlation decreases). In order to avoid unnecessary retransmissions, it is better to have a small amount of clusters under the responsability of a given RH when loss correlation is important.

6 Conclusion

In this paper, we have evaluated the impact of the tree structure on the retransmission efficiency for TRACK architectures. Our study is based on an analytical model using point processes in the plane and a loss model to represent the loss distributions of a multicast transport session. More precisely, our analytical model uses cluster processes which represents receiver concentration in local regions (such as AS, domains, etc.), and allows us to capture loss correlation amoung receivers. We have defined a cost function which represents transmission efficiency, as it counts the number of useless retransmissions of a data segment. Explicit formulae has been devised for the average cost function w.r.t generic receiver and RH distributions. These formulae allow us to deduce the optimal TRACK tree configurations. More precisely, we were able to give the optimal number of RHs which must be deployed, in order to maximize retransmission efficiency, with regards to specific topological parameters (such as loss probabilities, receiver distributions, etc.).

We are working on better describing the optimal structures (w.r.t. several cost functions) in order to get simple and generic dimensionning rules. We are also trying to collect precise statitistical information about loss distributions and receiver concentrations for multicast sessions, in order to get more realistic random geometric representation of the Internet.

References

1. S. Floyd *et al*. *A Reliable Multicast Framework for Light-weight Sessions and Application Level Framing*. In proceedings of SIGCOMM'95, ACM, October 1995.
2. T. Speakman *et al*. *PGM Reliable Transport Protocol Specification*. Internet Draft, IETF, April 2000.

3. *Reliable Multicast Transport Building Blocks for One-to-Many Bulk-Data Transfer* Request For Comments: 3048.
4. B. Whetten *et al. TRACK Architecture, A Scalable Real-Time Reliable Multicast Protocol.* Internet draft, IETF, November 2000.
5. B. Whetten *et al. Reliable Multicast Transport Building Block for TRACK.* IETF working document.
6. B.Whetten, G. Taskale, *Overview of the Reliable Multicast Transport Protocol II (RMTP-II).* IEEE Networking, Special Issue on Multicast, February 2000.
7. B. Neil Levine, JJ. Garcia-Luna-Aceves *A Comparison of Known Classes of Reliable Multicast Protocols*, in Proc. IEEE International Conference on Network Protocols, October 1996.
8. C. Maihöfer *A Bandwidth Analysis of Reliable Multicast Transport Protocols*, NGC 2000.
9. A. Busson, J-L Rougier, D. Kofman *Analysis and optimization of hierarchical reliable protocols* ITC 17.
10. A. Markopoulou, F. Tobagi *Hierarchical Reliable Multicast: performance analysis and placement of proxies* NGC 2000.
11. P. Bhagwat, P. Mishra, S. Tripathi *Effect on Topology on Performance of Reliable Multicast Communication* IEEE Infocom.
12. F.Baccelli, S.Zuyev. *Poisson-Voronoi Spanning Trees with applications to the Optimization of Communication Networks.* INRIA Research Report No. 3040. Nov.1996.
13. F. Baccelli, M. Klein, M. Lebourges and S. Zuyev, *Stochastic Geometry and Architecture of Communication Networks*, Telecommunication Systems, 7, pp. 209-227, 1997.
14. F.Baccelli, D.Kofman, J.L.Rougier.*Self-Organizing Hierarchical Multicast Trees and their Optimization.* IEEE Infocom'99, New-York (USA), March 1999.
15. RMT working group *Reliable Multicast Transport Building Block: Tree Auto-Configuration* draft-ietf-rmt-bb-tree-config-01.txt, IETF working document.
16. RMT working group *Reliable Multicast Transport Building Block: Tree Auto-Configuration*, draft-ietf-rmt-bb-tree-config-02.txt, IETF working document.
17. A. Chaintreau, C. Diot, F. Baccelli. *Impact of Network Delay on Multicast Sessions Performance With TCP-like Congestion Control*, Research Report No. 3987, INRIA Rocquencourt, September 2000.
18. N.G.Duffield, J.Horowotz, F.Lo Presti. *Adaptive Multicast Topology Inference.* in Proceedings of IEEE Infocom 2001, April 2001.
19. T.Friedman, D.Towsley. *Multicast Session Size Estimation.* in Proceedings of Infocom'99, March 1999.
20. B.N.Levine, J.Crowcroft, C.Diot, J.J.Garcia-Luna-Aceves, J.Kurose. *Consideration of Receiver Interest for IP Multicast Delivery.* in Proceedings of IEEE Infocom'00, March 2000.
21. F. Baccelli, K. Tchoumatchenko, S.Zuyev. Markov Paths on the Poisson-Delaunay Graph with applications to routing in mobile networks. INRIA. Rapport de recherche, 3420. To appear in Adv. Appl. Prob.
22. M. Yajnik, J. Kurose, and D. Towsley. *Packet Loss Correlation in the MBone Multicast Network.* IEEE Global Internet Conference 1996.
23. D.Stoyan, W.Kendall, J.Mecke. *Stochastic Geometry and its Applications.* 2nd Edition, J.Wiley & Sons Chichester, 1995.
24. D.Daley, D.Vere-Jones. *An Introduction to the Theory of Point Processes.* Springer Series in Statistics, Springer-Verlag New-York, 1988.
25. A. Okabe *et al. Spatial tesselations.* 2nd Edition, Wiley.

Framework for Authentication and Access Control of Client-Server Group Communication Systems*

Yair Amir, Cristina Nita-Rotaru, and Jonathan R. Stanton

Department of Computer Science, Johns Hopkins University
3400 North Charles St., Baltimore, MD 21218 USA
{yairamir,crisn,jonathan}@cs.jhu.edu

Abstract. Researchers have made much progress in designing secure and scalable protocols to provide specific security services, such as data secrecy, data integrity, entity authentication and access control, to multicast and group applications. However, less emphasis has been put on how to integrate security protocols with modern, highly efficient group communication systems and what issues arise in such secure group communication systems. In this paper, we present a flexible and modular architecture for integrating many different authentication and access control policies and protocols with an existing group communication system, while allowing applications to provide their own protocols and control the policies. This architecture maintains, as much as possible, the scalability and performance characteristics of the unsecure system. We discuss some of the challenges when designing such a framework and show its implementation in the Spread wide-area group communication toolkit.

1 Introduction

The Internet is used today not only as a global information resource, but also to support collaborative applications such as voice- and video-conferencing, whiteboards, distributed simulations, games and replicated servers of all types. Such collaborative applications often require secure message dissemination to a group and efficient synchronization mechanisms. Secure group communication systems provide these services and simplify application development.

A secure group communication system needs to provide confidentiality and integrity of client data, integrity, and possibly confidentiality, of server control data, client authentication, message source authentication and access control of system resources and services.

Many protocols, policy languages and algorithms have been developed to provide security services to groups. However, there has not been enough study of the integration of these techniques into group communication systems. Needed is a scheme flexible enough to accommodate a range of options and yet simple

---

* This work was supported by grant F30602-00-2-0526 from The Defense Advanced Research Projects Agency.

J. Crowcroft and M. Hofmann (Eds.): NGC 2001, LNCS 2233, pp. 128–140, 2001.

and efficient enough to appeal to application developers. Complete secure group communication systems are very rare and research on how to transition protocols into complete systems has been scarce.

Secure group systems really involve the intersection of three major, and distinct, research areas: networking protocols, distributed algorithms and systems, and cryptographic security protocols.

A simplistic approach when building a secure group system is to select a specific key management protocol, a standard encryption algorithm, and an existing access control policy language and integrate them with a messaging system. This would produce a working system, but would be complex, fixed in abilities, and hard to maintain as security features would be mixed with networking protocols and distributed algorithms.

In contrast, a more sophisticated approach is to construct an architecture that allows applications to plug-in both their desired security policy *and* the mechanisms to enforce the policy. Since each application has its particular security policies, it is natural to give an application more control not only on specifying the policy, but on the implementation of the services part of the policy too.

This paper proposes a new approach to group communication system architecture. More precisely, it provides such an architecture for authentication and access control. The architecture is flexible, allowing many different protocols to be supported and even be executing at the same time; it is modular so that security protocols can be implemented and maintained independently of the network and distributed protocols that make up the group messaging system; it allows applications to control what security services and protocols they use and configure; it efficiently enforces the chosen security policy without unduely impacting the messaging performance of the system.

As many group communication systems are built around a client-server architecture where a relatively small number of servers provide group communication services to numerous clients, we focused on systems utilizing this architecture. [1]

We implemented the framework in the Spread wide-area group communication system. We evaluate the flexibility and simplicity of the framework through six case studies of different authentication and access control methods. We show how both simple (IP based access control, password based authentication) and sophisticated (SecurID, PAM, anonymous payment, and group based) protocols can be supported by our framework.

Note that this paper is *not* a defense of any particular access control policy, authentication method or group trust model. Instead, it provides a flexible, complete interface to allow many such polices, methods, or models to be expressed and enforced by an existing, actively used group communication system.

The rest of the paper is organized as follows. Section 2 overviews related work. We present the authentication and access control framework and its implementation in the Spread toolkit in Section 3. We provide several brief case

---

[1] Some of the work may apply to network level multicast, but we have not explored that.

studies of how diverse protocols and policies can be supported by the framework in Section 4. Finally, we conclude and discuss future directions.

2   Related Work

There are two major directions in secure group communication research. The first one aims to provide security services for IP-Multicast and reliable IP-Multicast. Research in this area assumes a model consisting of one sender and many receivers and focuses on the high scalability of the protocols. Since the presence of a shared secret can be used as a foundation of efficiently providing data confidentiality and data integrity, a lot of work has been done in designing very scalable key management protocols. For lack of space we cite only the very recent ones: the VersaKey Framework [10] and the Group Secure Association Key Management Protocol (GSAKMP) [12].

The second major direction in secure group communication research is securing application level multicast systems, also known as group communication systems. These systems assume a many-to-many communication model where each member of the group can be both a receiver and a sender, and provide reliability, strong message ordering and group membership guarantees, with moderate scalability. Initially group communication systems were designed as high-availability, fault-tolerant systems, for use in local area networks. Therefore, the first group communication systems ISIS [9], Horus [21], Transis [4], Totem [5], and RMP [25] were less concerned with addressing security issues, and focused more on the ordering and synchronization semantics provided to the application (the Virtual Synchrony [8] and Extended Virtual Synchrony [17] models).

The number of secure group communication systems is small. Besides our system (Spread), the only implementation of group communication systems that focus on security are the RAMPART system at AT&T [20], the SecureRing [14] project at UCSB and the Horus/Ensemble work at Cornell [22]. A special case is the Antigone [16] framework, designed to provide mechanisms allowing flexible application security policies. Most relevant to this work are the Ensemble and the Antigone systems. Ensemble focused on optimizing group key distribution, and chose to allow application-dependent trust models in the form of access control lists treated as replicated data within the group. Authentication is achieved by using PGP. Antigone instead, allows flexible application security policies (rekeying policy, membership awareness policy, process failure policy and access control policy). However, it uses a fixed protocol to authenticate a new member and negotiate a key, while access control is performed based on a pre-configured access control list.

We also consider frameworks designed with the purpose of providing authentication and/or access control, without addressing group communication issues. Therefore, they are complementary to our work. One of these frameworks is the Pluggable Authentication Module (PAM) [23] which provides authentication services to UNIX system services (like login, ftp, etc). PAM allows an application not only to choose how to authenticate users, but also to switch dynamically between the authentication mechanisms without (rewriting and) recompiling a PAM-aware application. Other frameworks providing access control and authentication services are systems such as Kerberos [15] and Akenti [24]. Both of them have in common the idea of authenticating users and allowing access to resources, with the difference being that Kerberos uses symmetric cryptography, while Akenti uses public-key cryptography to achieve their goals.

One flexible module system that supports various security protocols is Flexinet [13]. Flexinet is an object oriented framework that focuses on dynamic negotiations, but does not provide any group-oriented semantics or services.

3   General System Architecture

The overall goal of this work is to provide a framework that integrates many different security protocols and supports all types of applications which have changing authentication and access control policy requirements, while maintaining a clear separation of the security policy from the group messaging system implementation. In this section, after discussing some design considerations, we present the authentication and access control frameworks.

3.1   Why Is a General Framework Needed?

When a communication system may only be used with one particular application, integrating the specific security policy and needed protocols with the system may make sense. However, when a communication system needs to support many different applications that may not always be cooperative, separating the policy issues which will be unique to each application from the enforcement mechanisms which must work for all applications avoids an unworkable "one-size-fits-all" security model, while maintaining efficiency.

Separating the policy implementation from both the application and the group communication system is also useful because in a live, production environment, the policy restrictions and access rules will change much more often than the code or system changes. So modifications of policy modules should not require recompiling or changing the application code.

The features of the general framework, as opposed to the features of a particular authentication or access control protocol, are:

1. Individual policies for each application.
2. Efficient policy enforcement in the messaging system.
3. Simple interface for both authentication and access control modules.
4. Independence of the messaging system from security protocols.
5. Many policies and protocols work with the framework, including: access control lists, password authentication, public/private key, certificates, role based access control, anonymous users, and dynamic peer-group policies.

We distinguish between authentication and access control modules to provide more flexibility. Each type of module has a distinctive interface which supports its specific task. The authentication module verifies that a client is who it claims to be. The access control module decides about all of the group communication specific actions a client attempts after it has been authenticated: join or leave a group, send an unicast message to another client or multicast a message to a group. It also decides whether a client is allowed to connect to a server (the access control module can deny a connection even if the authentication succeeded).

The framework supports *dynamic* policies. The main challenge with such policies is to allow changes during execution. Since the framework itself does not have any knowledge of the actual policy, for example it does not cache decisions or restrict what form actual policies take, it is possible for the access control modules to change how they make decisions independently of server. The modules need to make sure they activate dynamic changes in a consistent way, by using synchronized clocks, or by using the group communication services to agree on when to activate changes.

3.2   Framework Implementation in Spread

We implemented the framework in the Spread group communication system to give a concrete, real-world basis for evaluating the usefulness of this general architecture. Although we only implemented the framework within the Spread system, the model and the interface of the framework are actually quite general and the set of events upon which access control decisions can be made includes all of the available actions in a group-based messaging service (join, leave, group send, unicast send, connect).

3.3   The Spread Group Communication Toolkit

Spread [7,3,2] is a local and wide-area messaging infrastructure supporting reliable multicast and group communication. It provides reliability and ordering of messages (FIFO, causal, total ordering) and a membership service. The toolkit supports four different semantics: No membership, Closely Synchronous[2], Extended Virtual Synchrony (EVS) [1] and View Synchrony (VS) [11].

The system consists of one or more servers and a library linked with the application. The servers maintain most of the state of the system and provide reliable multicast dissemination, ordering of messages and the membership services. The library provides an API and basic services for message oriented applications. The application and the library can run on the same machine as a Spread server, in which case they communicate over IPC, or on separate machines, in which case the client-server protocol runs over TCP/IP.

Note that in order to implement our framework, we needed to modify both the Spread client library and the Spread daemon. When an application implements its own authentication and access control method, it needs to implement both the client side and the server side modules, however, it does not need to modify the Spread library or the Spread daemon.

---

[2] This is a relaxed version of EVS for reliable and FIFO messages.

In Spread each member of the group can be both a sender and a receiver. The system is designed to support small to medium size groups, but can accommodate a large number of different collaboration sessions, each of which spans the Internet. This is achieved by using unicast messages over the wide-area network and routing them between Spread nodes on an overlay network. Spread scales well with the number of groups used by the application without imposing any overhead on the network routers. Group naming and addressing is not a shared resource (as in IP multicast addressing), but rather a large space of strings which is unique to a collaboration session.

The Spread toolkit is available publicly and is being used by several organizations for both research and practical projects. The toolkit supports cross-platform applications and has been ported to several Unix platforms as well as Windows and Java environments.

3.4   Authentication Framework

All clients are authenticated when connecting to a server, and trusted afterwards. Therefore, when a client attempts actions, such as sending messages or joining groups, no authentication is needed. However, the attempted user actions are checked against a specified policy which controls which actions are permitted or denied for that user. This approach explicitly assumes that as long as a connection to the server is maintained, the same user is authenticated.

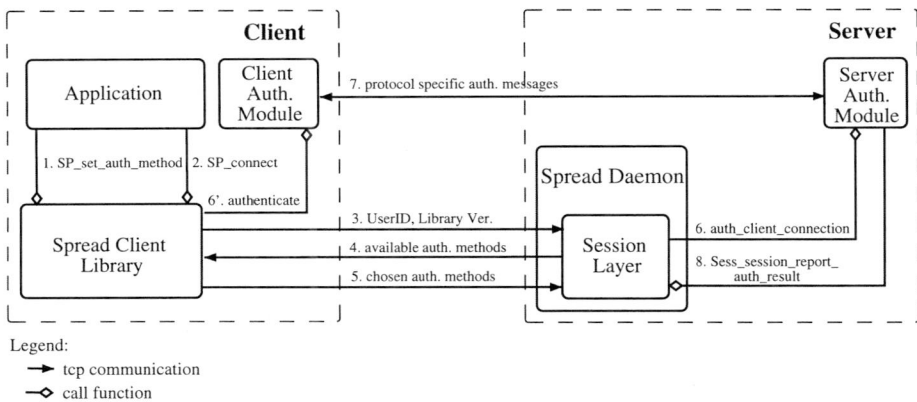

Fig. 1. Authentication Architecture and Communication Flow.

Figure 1 presents the architecture and the process of authentication. Both the client and the server implement an authentication module.

The change on the client side consists of the addition of a function (see Figure 2) that allows an application to set the authentication protocol it wishes

to use and to pass in any necessary data to that protocol, before connecting
to a Spread server. When the function that specifies the request of a client to
connect to a server is called (SP_connect), the connection tries to use the method
the application set to establish a connection. The authentication method chosen
by the application applies to all connections established by this application.

```
int SP_set_auth_method( const char *auth_name, int (*authenticate)
 (int, void *), void * auth_data );
int SP_set_auth_methods( int num_methods, const char *auth_name[],
 int (*authenticate[]) (int, void *), void * auth_data[] );

/* declaration of authenticate function */
int authenticate(int fd, void * user_data_pointer);
```

Fig. 2. Client Authentication Module API.

A server authentication module needs to implement the functions listed in
the auth_ops structure (see Figure 3, line 10). Then the module should register
itself with the Spread daemon by calling the Acm_auth_add_method function. By
default, a module is registered in the 'disabled' state. The system administrator
can enable the module when configuring Spread.

The authentication process begins when the session layer of the daemon receives a connection request from a client. After some initial information exchange
and negotiation of the allowed authentication protocols, the session module constructs a session_auth_info structure containing the list of agreed upon authentication protocols. This structure is passed as a parameter to each authentication
function and is used as a handle for the entire process of authenticating a client.
The authentication function can use the module_data pointer to store any module specific data that it needs during authentication. The session layer calls the
auth_client_connection method for each protocol and then "forgets about" the
client connection. A minimal state about the client is stored, but no messages
are received or delivered to the client at this point.

The auth_client_connection function is responsible for authenticating the
client connection. If authenticating the client will take a substantial amount of
CPU or real time, the function should not do the work directly, but rather setup
a callback function to be called later (for example when messages arrive from
the client), and then it should return. Another approach is to fork off another
process to handle the authentication. This is required because the daemon is
blocked while this function is running.

The auth_client_connection function never returns a decision value because
a decision may not have been reached yet. When a decision has been made the
server authentication module calls Sess_session_report_auth_result and releases
control to the session layer. The Sess_session_report_auth_result function re-
ports whether the current authentication module has successfully authenticated

```
struct session_auth_info {
    int ses;
    void *module_data;
    int num_required_auths;
    int completed_required_auths;
    int required_auth_methods[MAX_AUTH_METHODS];
    int required_auth_results[MAX_AUTH_METHODS];
};

struct auth_ops {
    void (*auth_client_connection) (struct session_auth_info
    *sess_auth_p);
};

struct acp_ops {
    bool (*open_connection) (char *user);
    bool (*open_monitor) (char *user); /* not used currently */
    bool (*join_group) (char *user, char *group, void *acm_token);
    bool (*leave_group) (char *user, char *group, void *acm_token);
    bool (*p2p_send) (char *user, char dests[][MAX_GROUP_NAME],
      int service_type);
    bool (*mcast_send) (char *user, char groups[][MAX_GROUP_NAME],
      int service_type);
};

/* Auth Functions */
bool Acm_auth_add_method(char *name, struct auth_ops *ops);

/* Access Control Policy Functions */
bool Acm_acp_set_policy(char *policy_name);
bool Acm_acp_add_method(char *name, struct acp_ops *ops);
```

Fig. 3. Server Authentication and Access Control Module API.

the session or not. If more than one authentication method was required, the connection succeeds if all the methods succeed.

3.5   Access Control Framework

In our model, an authenticated client connection is not automatically allowed to perform any actions. Each action a client may request of the server, such as sending a message or joining or leaving a group, is checked at the time it is attempted against an access control policy module. The enforcement checks are implemented by having the session layer of the server call the appropriate access control policy module callback function (see Figure 3, lines 14-20) return a decision. The implementation of the check functions should be optimized as they have a direct impact on the performance of the system as they are called for every client action.

If the module chooses to allow the request, then the server handles it normally. In the case of rejection, the server creates a special "reject" message which will be sent to the client in the normal stream of messages. The reject message contains as much of the data included in the original attempt as possible. The application should be able to identify which message was rejected by whatever information it stored in the body of the message (such as an application level sequence number) and respond to it appropriately. That response could be a notification to the user, establishing a new connection with different authentication credentials and retrying the request, logging an error, etc.

The server can reject an action at two points, when the server receives the action from the client or when the action is going to take effect. For example, when a client joins a group the join can be rejected when the join request is received from the directly connected client, and when the join request has been sent to all of the servers and has been totally ordered. Rejecting the request the first time it is seen avoids processing requests that will later be rejected and simplifies the decision-making because only the server the client is directly connected to will make the decision. The disadvantage is that at the time the request is being accepted or rejected the module only knows the current state of the group or system and not what the state will be when the request would be acted upon by the servers. Since these states can differ, some type of decisions may not be possible at the early decision point.

4   Case Studies

To provide some intuition as to what building a Spread authentication module requires, this section discusses the implementation of several real-world modules: an IP based access control module, a password based authentication module, a SecurID or PAM authentication module, an anonymous payment authentication and anonymous access control module, and a dynamic peer-group authentication module. For more details and implementation code see [6].

IP Access Control. A very simple access control method that does not involve any interaction with the client process or library, is one that is based on the IP address of the clients. The connection is allowed considering the IP address from which the client connected to the server. This module only restricts the open_connection (see Figure 3, line 15) operation.

Password Authentication. A common form of authentication uses some type of password and username to establish the identity of the user. Many types of password based authentication can be supported by our framework from passwords sent in the clear (like in telnet) to challenge-response passwords.

To implement a password-based authentication method, both a client and a server side need to be implemented. The server module can use the Events subsystem in Spread to wait for network events to occur and avoid blocking the server while the user is entering its password or the client and server modules are communicating. The client module consists of one function which is called during the establishment of a connection and returns either success or failure. The function can use the file descriptor of the socket over which the connection is being established and whatever data pointer was registered by the SP_set_auth_method. In this case the application prompted the user for a username and password and created a user_password structure. The authenticate function, sends the username and the password to the server and waits for a response, informing it of whether or not the authentication succeeded.

SecurID. A popular authentication method is RSA SecurID. The method uses a SecurID server to authenticate a SecurID client based on a unique randomly generated identifier and a PIN. In some cases the SecurID server might ask the client to provide new credentials. We do not discuss here the internal of the SecurID authentication mechanism (see [19] for more details), but focus on how our framework can accommodate this method.

The main difference from the previous examples is that here the Server Authentication Module needs to communicate with the SecurID server. As mentioned before, the auth_client_connection function should not block. Blocking can happen when opening a connection with a SecurID server and retrieving messages from it. Therefore, auth_client_connection forks another process responsible for the authentication protocol and then registers an event such that it will get notified when the forked process finished. The forked process establishes a connection with the SecurID Server and authenticates the user. When it finishes, the Server Authentication Module gets notified, so it can call the Sess_session_report_auth_result function to inform the Spread daemon that a decision was taken and to pass control back to it.

PAM. Another popular method of authentication is the modular PAM [23] system which is standard on Solaris and many Linux systems. Here the authentication module will act as a client to a PAM system and request authentication through the standard PAM function calls. To make authentication through PAM work, the module must provide a way for PAM to communicate and interact with the actual human user of the system, to prompt for a password or other information. The module would register an interactivity function with PAM that would pass all of the requests to write to the user or request input from the user over the Spread communication socket to the Spread client authentication module for PAM. This client module would then act on the PAM requests and interact with the user and then send the reply back to the Spread authentication module which would return the results to the actual PAM function.

Anonymous Payments. An interesting approach is when access is provided to anonymous clients in exchange for payment. These systems [18] perform transactions between a client and a merchant, assuming that both of them have accounts with a Bank. By using cryptographic techniques, the system provides anonymity of the client and basic security services. We do not detail the cryptographic details, but show how this method can be accommodated in our framework.

We assume support from the anonymous payments system (in the form of an API) and require the servers and the client to have an account with a Bank. When a client connects to a server, the Client Authentication Module generates a check and an identifier of client's account and then passes them to the Server Authentication Module which will then contact the Bank to validate the check (if necessary another process will be forked as in the SecurID case). When validated, the Server Authenticated Module will register the client's identifier with the access control policy as a paid user of the appropriate groups. Then, for as long as the payment was valid, the client will be permitted to access the groups they paid for and the server has no knowledge of the client's identity.

Group-Based Authentication. In all the previous authentication methods presented, the authentication of a client is handled by the server that the client connects to. In larger, non-homogeneous environments authentication may involve some or all of the group communication system servers. Although these protocols may be more complex, they can provide better mappings of administrative domains, and possibly better scalability.

An example of such a protocol is when a server does not have sufficient knowledge to check a client's credentials (for instance a certificate). In this case, it sends the credentials to all the servers in the configuration and each server then attempts to check the credentials itself and sends an answer back. If at least one server succeeds, the client is authenticated. The particularity of such a protocol is that the servers need to communicate between them as part of the authentication process. Since all the servers can communicate between them in our system, the framework provides all necessary features that allows the integration of such a group-based authentication method.

Access Control. We realize that the above case studies are focused on authentication. Few standard access control protocols that we could use as case studies exist. To demonstrate the ability of the access control architecture we create a case study about an imaginary secure IRC system. Consider a set of users where some users are allowed to chat on the intelligence group, while others are restricted to the operations group. Some are allowed to multicast to a group but are not allowed to read the group messages (virtual drop-box). Our framework supports these access control policies through appropriate implementation of the join and multicast hooks defined in Figure 3. Access control modules support identity based, role based, or credential based restrictions.

5   Conclusions and Future Work

We presented a flexible implementation of an authentication and access control framework in the Spread wide area group communication system. Our approach allows an application to write its own authentication and access control modules, without needing to modify the Spread client or server code. The flexibility of the system was showed by showing how a wide range of authentication methods can be implemented in our framework.

There are a lot of open problems that are subject of future work. These include: providing tools that allow an application to actually specify a policy, handling policies in a system supporting network partitions (for example merging components with different policies), providing support for meta-policies defining which entity is allowed to create or modify group policies, and developing dynamic group trust protocols for authentication.

References

1. Y. Amir. *Replication using Group Communication over a Partitioned Network.* PhD thesis, Institute of Computer Science, The Hebrew University of Jerusalem, Jerusalem, Israel, 1995.
2. Y. Amir, B. Awerbuch, C. Danilov, and J. Stanton. Flow control for many-to-many multicast: A cost-benefit approach. Technical Report CNDS-2001-1, Johns Hopkins University, Center of Networking and Distributed Systems, 2001.
3. Y. Amir, C. Danilov, and J. Stanton. A low latency, loss tolerant architecture and protocol for wide area group communication. In *Proceedings of the International Conference on Dependable Systems and Networks*, pages 327–336, June 2000.
4. Y. Amir, D. Dolev, S. Kramer, and D. Malki. Transis: A communication subsystem for high availability. *Digest of Papers, The 22nd International Symposium on Fault-Tolerant Computing Systems*, pages 76–84, 1992.
5. Y. Amir, L. E. Moser, P. M. Melliar-Smith, D. Agarwal, and P. Ciarfella. The totem single-ring ordering and membership protocol. *ACM Transactions on Computer Systems*, 13(4):311–342, November 1995.
6. Y. Amir, C. Nita-Rotaru, and J. Stanton. Framework for authentication and access control of client-server group communication systems. Technical Report CNDS 2001-2, Johns Hopkins University, Center of Networking and Distributed Systems, 2001. `http://www.cnds.jhu.edu/publications/`.
7. Y. Amir and J. Stanton. The Spread wide area group communication system. Technical Report 98-4, Johns Hopkins University, Center of Networking and Distributed Systems, 1998.
8. K. P. Birman and T. Joseph. Exploiting virtual synchrony in distributed systems. In *11th Annual Symposium on Operating Systems Principles*, pages 123–138, November 1987.
9. K. P. Birman and R. V. Renesse. *Reliable Distributed Computing with the Isis Toolkit.* IEEE Computer Society Press, March 1994.
10. G. Caronni, M. Waldvogel, D. Sun, N. Weiler, and B. Plattner. The VersaKey framework: Versatile group key management. *IEEE Journal of Selected Areas in Communication*, 17(9), September 1999.
11. A. Fekete, N. Lynch, and A. Shvartsman. Specifying and using a partitionable group communication service. In *Proceedings of the 16th annual ACM Symposium on Principles of Distributed Computing*, pages 53–62, Santa Barbara, CA, August 1997.
12. H. Harney, A. Colegrove, E. Harder, U. Meth, and R. Fleischer. Group secure association key management protocol (GSAKMP). draft-irtf-smug-gsakmp-00.txt, November 2000.

13. R. Hayton, A. Herbert, and D. Donaldson. FlexiNet — A flexible component oriented middleware system. In *Proceedings of SIGOPS'98*, http://www.ansa.co.uk/, 1998.
14. K. P. Kihlstrom, L. E. Moser, and P. M. Melliar-Smith. The SecureRing protocols for securing group communication. In *Proceedings of the IEEE 31st Hawaii International Conference on System Sciences*, volume 3, pages 317–326, Kona, Hawaii, January 1998.
15. J. Kohl and B. C. Neuman. The Kerberos Network Authentication Service (Version 5). RFC-1510, September 1993.
16. P. McDaniel, A. Prakash, and P. Honeyman. Antigone: A flexible framework for secure group communication. In *Proceedings of the 8th USENIX Security Symposium*, pages 99–114, August 1999.
17. L. E. Moser, Y. Amir, P. M. Melliar-Smith, and D. A. Agarwal. Extended virtual synchrony. In *Proceedings of the IEEE 14th International Conference on Distributed Computing Systems*, pages 56–65. IEEE Computer Society Press, Los Alamitos, CA, June 1994.
18. B. C. Neuman and G. Medvinsky. Requirements for network payment: The netcheque perspective. In *In Proceedings of IEEE COMPCON'95*, March 1995.
19. M. Nystrom. The SecurID SASL mechanism. RFC-2808, April 2000.
20. M. K. Reiter. Secure agreement protocols: Reliable and atomic group multicast in RAMPART. In *Proceedings of the 2nd ACM Conference on Computer and Communications Security*, pages 68–80. ACM, November 1994.
21. R. V. Renesse, K.Birman, and S. Maffeis. Horus: A flexible group communication system. *Communications of the ACM*, 39:76–83, April 1996.
22. O. Rodeh, K. Birman, and D. Dolev. The architecture and performance of security protocols in the Ensemble Group Communication System. *ACM Transactions on Information and System Security*, To appear.
23. V. Samar and R. Schemers. Unified login with Pluggable Authentication Modules (PAM). OSF-RFC 86.0, October 1995.
24. M. Thompson, W. Johnston, S. Mudumbai, G. Hoo, K. Jackson, and A. Essiari. Certificate-based access control for widely distributed resources. In *Proceedings of the Eighth Usenix Security Symposium*, pages 215–227, August 1999.
25. B. Whetten, T. Montgomery, and S. Kaplan. A high performance totally ordered multicast protocol. In *Theory and Practice in Distributed Systems, International Workshop*, Lecture Notes in Computer Science, page 938, September 1994.

Scalable IP Multicast Sender Access Control for Bi-directional Trees

Ning Wang and George Pavlou

Center for Communication Systems Research, University of Surrey, United Kingdom
{N.Wang,G.Pavlou}@eim.surrey.ac.uk

Abstract. Bi-directional shared tree is an efficient routing scheme for interactive multicast applications with multiple sources. Given the open-group IP multicast service model, it is important to perform sender access control so as to prevent group members from receiving irrelevant data, and also protect the multicast tree from various Denial-of-Service (DoS) attacks. Compared with source specific trees and uni-directional shared trees where information sources can be authorized or authenticated at the single root or Rendezvous Point (RP), in bi-directional trees this problem becomes challengeable since hosts can send data to the shared tree from any network point. In this paper we propose a scalable sender access policy mechanism for bi-directional shared trees so that irrelevant data is policed and discarded once it hits any on-tree router. We consider the scenario of both intra-domain and inter-domain routing in the deployment of the policy, so that the mechanism can adapt to situations in which large-scale multicast applications or many concurrent multicast sessions are involved, potentially across administrative domains.

1 Introduction

IP multicast [9] supports efficient communication services for applications in which an information source sends data to a group of receivers simultaneously. Although some IP multicast applications have been available on the experimental Multicast Backbone (*MBone*) for several years, large-scale deployment has not been achieved until now. IP multicast is also known as "Any Source Multicast (*ASM*)" in that an information source can send data to any group without any control mechanism. In the current service model, group management is not stringent enough to control both senders and receivers. *IGMPv2* [11] is used to manage group members when they join or leave the session but in this protocol there are no control mechanisms to avoid receiving data from particular information sources or prevent particular receivers from receiving sensitive information. It has been observed that the above characteristics of IP multicast have somehow prevented successful deployment of related applications at large scale on the Internet [10].

Realizing that many multicast applications are based on one-to-many communications, e.g. Internet TV/radio, pushed media, etc., H. W. Holbrook et al proposed the *EXPRESS* routing scheme [14], from which the Source Specific Multicast (*SSM*) [15] service model was subsequently evolved. In *SSM* each group is

J. Crowcroft and M. Hofman (Eds.): NGC 2001, LNCS 2233, pp. 141-158, 2001.
©Springer-Verlag Berlin Heidelberg 2001

identified by an address tuple (S, G) where S is the unique address of the information source and G is the destination channel address. A single multicast tree is built rooted at the well-known source for delivering data to all subscribers. Under such a scenario, centralized group authorization and authentication can be achieved at the root of the single source at the application level. Currently *IGMPv3* [7] is under the development to support source specific joins in *SSM*.

On the other hand, it should be noted that there exist many other applications based on many-to-many styled communication, such as multi-party videoconferencing system, Distributed Interactive Simulation (*DIS*) and Internet games etc. For this type of interactive applications, bi-directional multicast trees such as Core Based Tree (*CBT*) [2], Bi-directional *PIM* [13], and *RAMA* style Simple Multicast [19], are efficient routing schemes for natively delivering data between multiple hosts. However, since there is no single point for centralized group access control, sender authorization and authentication become new challenges. Typically, if a malicious host wants to perform Denial-of-Service (*DoS*) attack it can flood bogus data from any point of the bi-directional multicast tree. Sender access control for bi-directional trees based on IP multicast model is not provided in the specification of any corresponding routing protocols such as [2, 13]. One possible solution that has been proposed is to periodically "push" the entire sender access list down to all the on-tree routers, so that only data from authorized senders can be accepted and sent onto the bi-directional tree [6]. This simple access control mechanism has been adopted in the *RAMA*-style Simple Multicast [14]. However, this policy is not very scalable especially when many multicast groups or large group size with many senders are involved. A more sophisticated scheme named *Keyed-HIP* (*KHIP*) [21] works on the routing level to provide data access control on the bi-directional tree, and flooding attacks can be also detected and avoided by this network-level security routing scheme.

In this paper we will propose an efficient and scalable sender access control mechanism for bi-directional trees in the IP multicast service model. The basic idea is to deploy access policy for external senders on the tree routers where necessary, so that data packets from unauthorized senders will be policed and discarded once it hits the bi-directional tree. Our proposed scheme causes little impact on the current bi-directional routing protocols so that it can be directly implemented on the Internet without modifying the basic function of the current routing protocols. Moreover, the overhead introduced by the new control mechanism is much smaller than that proposed in [6] and [19].

The rest of the paper is organized as follows: Section 2 gives the overview of our proposed dynamic maintenance of the policy. Sections 3 and 4 introduce sender authorization and authentication in intra-domain and inter-domain routing. Operations on multi-access networks are specially discussed in section 5. We examine the scalability issues of our proposed scheme in section 6, and finally we present a summary in section 7.

2   Sender Authorization and Authentication Overview

Compared with source specific trees and even uni-directional shared trees such as *PIM-SM* [8], in which external source filtering can be performed at the single source or Rendezvous Point (*RP*) where the registrations of all the senders are processed and authorized, in bi-directional trees this is much more difficult since data from any source will be directly forwarded to the whole tree once it hits the first on-tree router. In fact, since there is no single point for centralized sender access control, information source authorization and authentication has to be deployed at the routing level. As we have already mentioned, the simplest solution for this is to periodically broadcast the entire access control list down to all the routers on the bi-directional tree for deciding whether or not to accept data (e.g., [19]). However, this method is only feasible when a few small-sized groups with limited number of senders are considered. For large scale multicast applications, if we don't send the whole policy down to all the on-tree routers so as to retain the scalability, three questions need to be answered as proposed in [6]: (1) How to efficiently distribute the list where necessary? (2) How to find edge routers that act as the trust boundary? (3) How to avoid constant lookups for new sources? In fact if we try to statically mount the access control policy to an *existing* bi-directional multicast tree, none of the above three questions can be easily answered.

It should be noted that most multicast applications are highly dynamic by nature, with frequent join/leaving of group members and even information senders. Hence the corresponding control policy should also be dynamically managed. Here we propose an efficient sender-initiated distribution mechanism of the access list during the phase of multicast tree construction. The key idea is that each on-tree router only adds its *downstream* senders to the local Sender Access Control List (*SACL*) during their join procedure, and the senders in the access list are activated by the notification from the core. In fact, only the core has the right to decide whether or not to accept the sources and it also maintains the entire *SACL* for all the authorized senders. Packets coming from any unauthorized host (even if it has already been in the tree) will be discarded at once when they reach any on-tree router. To achieve this, all senders must first register with the core before they can send any data to the group. When a registration packet hits an on-tree router, the unicast address of the sender is added into the *SACL* of each router on the way. Under this scenario, the access policy for a particular sender is deployed on the branch from the first on-tree router where the registration is received along to the core router. Here we define the interface from which this registration packet is received as the *downstream interface* and the one used to deliver unicast data to the core as the *upstream interface*. The format of each *SACL* entry is (*G, S, I*) where *G* indicates the group address, *S* identifies the sender and *I* is the downstream interface from which the corresponding registration packet was received. If the core has approved the join, it will send a type of "activating packet" back to the source, and once each on-tree router receives this packet, it will activate the source in its *SACL* so that it will be able to send data onto the bi-directional tree from then on. Under such a scenario, an activated source can only send group data to the tree via the path where its *SACL* entry has been recorded, i.e., even if a sender has been authorized, it cannot send data to the group from other branches or elsewhere. Source authentication entries kept in each *SACL* are maintained in soft state for flexibility purpose, and this requires that information sources should periodically send refreshing packets to the core to keep their states alive in the upstream routers. This action is especially necessary when a source is temporarily not sending group data. Once data packets have been received from a particular registered sender, the on-tree router may assume that this source is still alive and will automatically refresh the state for it. If a particular link between the data source and the core fails, the corresponding state will time out and become obsolete. In this case the host has to seek alternative path to perform re-registration for continuing sending group data.

When a router receives a data packet from one of its downstream interfaces, it will first check if there exists such an entry for the data source in its local *SACL*. If the router cannot find a matching entry that contains the unicast address of the source, the data packet is discarded. Otherwise if the corresponding entry has been found, the router will verify if this packet comes from the same interface as the one recorded in the *SACL* entry. Only if the data packet has passed these two mechanisms of authentication, it will be forwarded to the upstream interface and the other interfaces with the group state, i.e., interfaces where receivers are attached. On the other hand, when a data packet comes from the upstream interface, the router will always forward it to all the other interfaces with group state and need not perform any authentication. Although the router cannot judge if this data packet is from a registered sender, since it comes from the upstream router, there exist only two possibilities: either the upstream router has the *SACL* entry for the data source or the upstream router has received the packet from its own parent router in the tree. The extreme case is that none of the intermediate ancestral routers have such an entry and then we have to backtrack to the core. Since the core has recorded entries for all the registered senders and it never forwards any unauthenticated packet on its downstream interfaces, we can safely conclude that each on-tree router can trust its parent, and hence packets received from the upstream interface are always from valid senders. However, this scenario precludes the case of routers attached on multi-access networks such as *LANs*, and we will discuss the corresponding operations in section 5.

Compared with source specific trees and even uni-directional shared trees such as *PIM-SM* [8], in which external source filtering can be performed at the single source or Rendezvous Point (*RP*) where the registrations of all the senders are processed and authorized, in bi-directional trees this is much more difficult since data from any source will be directly forwarded to the whole tree once it hits the first on-tree router. In fact, since there is no single point for centralized sender access control, information source authorization and authentication has to be deployed at the routing level. As we have already mentioned, the simplest solution for this is to periodically broadcast the entire access control list down to all the routers on the bi-directional tree for deciding whether or not to accept data (e.g., [19]). However, this method is only feasible when a few small-sized groups with limited number of senders are considered. For large scale multicast applications, if we don't send the whole policy down to all the on-tree routers so as to retain the scalability, three questions need to be answered as proposed in [6]: (1) How to efficiently distribute the list where necessary? (2) How to find edge routers that act as the trust boundary? (3) How to avoid constant lookups for new sources? In fact if we try to statically mount the access control policy to an *existing* bi-directional multicast tree, none of the above three questions can be easily answered.

It should be noted that most multicast applications are highly dynamic by nature, with frequent join/leaving of group members and even information senders. Hence the corresponding control policy should also be dynamically managed. Here we propose an efficient sender-initiated distribution mechanism of the access list during the phase of multicast tree construction. The key idea is that each on-tree router only adds its *downstream* senders to the local Sender Access Control List (*SACL*) during their join procedure, and the senders in the access list are activated by the notification from the core. In fact, only the core has the right to decide whether or not to accept the sources and it also maintains the entire *SACL* for all the authorized senders. Packets coming from any unauthorized host (even if it has already been in the tree) will be discarded at once when they reach any on-tree router. To achieve this, all senders must first register with the core before they can send any data to the group. When a registration packet hits an on-tree router, the unicast address of the sender is added into the *SACL* of each router on the way. Under this scenario, the access policy for a particular sender is deployed on the branch from the first on-tree router where the registration is received along to the core router. Here we define the interface from which this registration packet is received as the *downstream interface* and the one used to deliver unicast data to the core as the *upstream interface*. The format of each *SACL* entry is (*G, S, I*) where *G* indicates the group address, *S* identifies the sender and *I* is the downstream interface from which the corresponding registration packet was received. If the core has approved the join, it will send a type of "activating packet" back to the source, and once each on-tree router receives this packet, it will activate the source in its *SACL* so that it will be able to send data onto the bi-directional tree from then on. Under such a scenario, an activated source can only send group data to the tree via the path where its *SACL* entry has been recorded, i.e., even if a sender has been authorized, it cannot send data to the group from other branches or elsewhere. Source authentication entries kept in each *SACL* are maintained in soft state for flexibility purpose, and this requires that information sources should periodically send refreshing packets to the core to keep their states alive in the upstream routers. This action is especially necessary when a source is temporarily not sending group data. Once data packets have been received from a particular registered sender, the on-tree router may assume that this source is still alive and will automatically refresh the state for it. If a particular link between the data source and the core fails, the corresponding state will time out and become obsolete. In this case the host has to seek alternative path to perform re-registration for continuing sending group data.

When a router receives a data packet from one of its downstream interfaces, it will first check if there exists such an entry for the data source in its local *SACL*. If the router cannot find a matching entry that contains the unicast address of the source, the data packet is discarded. Otherwise if the corresponding entry has been found, the router will verify if this packet comes from the same interface as the one recorded in the *SACL* entry. Only if the data packet has passed these two mechanisms of authentication, it will be forwarded to the upstream interface and the other interfaces with the group state, i.e., interfaces where receivers are attached. On the other hand, when a data packet comes from the upstream interface, the router will always forward it to all the other interfaces with group state and need not perform any authentication. Although the router cannot judge if this data packet is from a registered sender, since it comes from the upstream router, there exist only two possibilities: either the upstream router has the *SACL* entry for the data source or the upstream router has received the packet from its own parent router in the tree. The extreme case is that none of the intermediate ancestral routers have such an entry and then we have to backtrack to the core. Since the core has recorded entries for all the registered senders and it never forwards any unauthenticated packet on its downstream interfaces, we can safely conclude that each on-tree router can trust its parent, and hence packets received from the upstream interface are always from valid senders. However, this scenario precludes the case of routers attached on multi-access networks such as *LANs*, and we will discuss the corresponding operations in section 5.

3   Intra-domain Access Control Policy

3.1   *SACL* Construction and Activation

As we have mentioned, all the information sources must register with the core before they can send any data to the bi-directional tree. For each on-tree router, its *SACL* is updated when the registration packet of a new sender is received, and the individual entry is activated when its corresponding activating notification is received from the core.

If a host wants to both send and receive group data, it must join the multicast group and become a Send-Receive capable member (SR-member, *SRM*). Otherwise if the host only wants to send messages to the group without receiving any data, it may choose to act as a Send-Only member (SO-member, *SOM*) or a Non-Member Sender (*NMS*). In the former case, the host must join the bi-directional tree to directly send the data, and its designated router will forward the packets on the upstream interface as well as other interfaces with the group state. In the IP multicast model information sources are allowed to send data to the group without becoming a member. Hence, if the host is not interested in the information from the group, it may also choose to act as a non-member sender. In this case, the host must encapsulate the data and unicast it towards the core. Once the data packet hits the first on-tree router and passes the corresponding source authentication, it is decapsulated and forwarded on all the other interfaces with the group state. The following description is based on the *CBT* routing protocol, but it can also apply to other bi-directional routing schemes such as Bidir-*PIM* and *RAMA*-style Simple Multicast.

(1)   SR-member Join

When the Designated Router (*DR*) receives a group *G* membership report from a SR-member *S* on the *LAN*, it will send a join request towards the core. Here we note that the group membership report cannot be suppressed by the *DR* if it is submitted by a send-capable member. Once a router receives this join-request packet from one of its interfaces, say, *A*, then the (*G, S, A*) entry is added into its *SACL*. If the router is not been on the shared tree, a (*, G*) state is created with the interface leading to the core as the upstream interface and *A* is set to the downstream interface. At the same time, interface *A* is also added to the interface list with group state so that data from other sources can be forwarded to *S* via *A*. If the router already has the (*, G*) state, but *A* is not in the interface list with group state, then it is added to the list. Thereafter, the router just forwards the join-request to the core via its upstream interface. Once the router receives the activating notification from the core, the (G, S, A) entry is activated so that S is able to send data.

(2) SO-member join

Similar to SR-member joins, the *DR* of a SO-member also sends a join-request up to the core and when the router receives this request from its interface A, the (G, S, A) entry is added to the local *SACL*. If the router is not yet on the tree, $(*, G)$ state will be generated but the interface A is not added to the interface list with group state. This is because A needs not to forward group data to a send-only member.

(3) Non-Member Sender (*NMS*) registration

Here we use the terminology "registration" instead of "join request", since this host is not a group member and need not be on the tree to send group data. The registration packet from the Non-Member Sender is unicast towards the core and when it hits the first router with $(*, G)$ state, the (G, S, A) entry will be created in the local *SACL* of all the on tree routers on the way leading to the core. It should be noted, if a router is not on the tree, it does not maintain *SACL* for the group.

Finally, if a receive-only member (also known as the *group member* in conventional multicast model) wants to join the group, the join request only invokes a $(*, G)$ state if the router is not on the tree, but no new *SACL* entries need to be created. Moreover, once the join request hits any on-tree router, a join-notification is immediately sent back without informing the core.

The forwarding behavior of an on-tree router under send access control mechanism is as follows. If group data comes from downstream interfaces, the router will authenticate the information source by looking up the local *SACL* and if the sender has its entry in the list and comes from the right interface, the data is forwarded on the upstream interface and other interfaces with group state. If the corresponding *SACL* check fails, the data is discarded at once. On the other hand, if the data comes from the upstream interface, it is forwarded to all the other interfaces with the group state because a router's parent is always trusted by its children.

3.2   An Example for Intra-domain Access Policy

A simple network model is given in Fig. 1 . We assume that node A is the core router and all the Designated Routers (*DR*) of potential members of group G should send join request to this node. Hosts *H1-H5* are attached to the individual routers as shown in the figure.

Initially suppose *H1* wants to join the group, its *DR* (router B) will create $(*, G)$ state and send the join request to the core A. Since *H1* is a SR-member that can both send and receive data to/from the group, each of the routers that the join request has passed will add this sender into its local *SACL*. Hence both router B and A will have the *SACL* entry $(G, H1, 1)$, since they both receive the join request from interface 1.

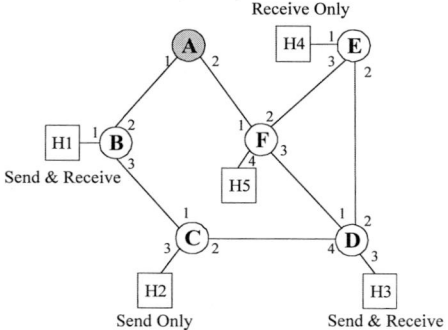

Fig. 1. Intra-domain Network Model.

Host *H2* only wants to send messages to group *G* but does not want to receive any data from this group, and so it may choose to join as a SO-member or just act as a *NMS*. In the first case, its *DR* (router *C*) will create (*, *G*) state indicating that this router is an on-tree node and then add *H2* to its *SACL*. Thereafter, router *C* will send a join request indicating *H2* is a SO-member towards the core; when *B* receives this request, it will also add *H2* to its local *SACL* and then forward the join-request packet to *A*. Since *H2* does not want to receive data from the group, link *BC* becomes a send-only branch. To achieve this, router *B* will not add *B3* to the interface list with group state. If *H2* chooses to act as the Non-Member Sender, router *C* will not create (*, *G*) state or *SACL* for the group but send a registration packet towards *A*. When this packet hits an on-tree router, say, *B* in our example, *H2* will be added to the local *SACL* of all the routers on the way. When sending group messages, router *C* just encapsulates the data destined to the core by setting the corresponding IP destination address to *A*. When the data reaches *B* and passes the *SACL* authentication, the IP destination address is changed to the group address originally contained in the option field of the data packet, and the message is forwarded to interfaces *B1* and *B2* to get to *H1* and the core respectively. After *H3* and *H4* join the group, the resulting shared tree is shown in Fig. 2, and *SACLs* of each on-tree router are also indicated in the figure. It should be noted that *H4* is a receive-only member, and hence Router *E*, *F* and *A* need not add it to their local *SACLs*. Suppose router *F* has received group data from *H3* on interface *F3*, it will check in its local *SACL* if *H3* is an authorized sender. When the data passes the address and interface authentications, it is forwarded to both interfaces *F1* and *F2*. When group data is received on the upstream interface *F1*, since its parent *A* is a trusted router (in fact the data source should be either *H1* or *H2*), the data is forwarded to *F2* and *F3* immediately without any authentication. However, if the non-registered host *H5* wants to send messages to the group, data won't be forwarded to the bi-directional tree due to the *SACL* authentication failure at router *F*.

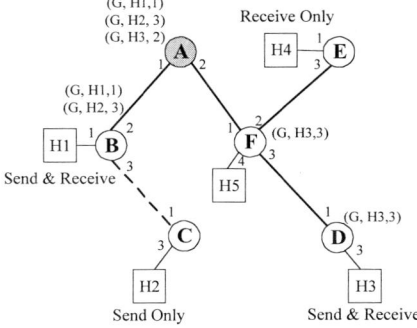

Fig. 2. Bi-directional Tree with *SACL*.

4   Inter-domain Access Control Policy

4.1   Basic Descriptions

As we have mentioned above, on-tree routers only maintain the access policy for all the downstream senders. However, if large-scale groups with many senders or many concurrent sessions are considered, the size of the *SACL* in the routers near the core will become a heavy burden for these on-tree routers. In this section we discuss how this situation can be improved with the aid of inter-domain IP multicast routing semantics.

Our key idea is based on hierarchical access control policy to achieve scalability. All routers only maintain *SACL* for the downstream senders in the *local* domain and need not add sources from downstream domains to their local *SACLs*. In other words, all the senders for the group are only authenticated in the local domain. In the root domain, the core needs to keep entries only for local senders; however in order to retain the function of authorizing and activating information sources from remote domains, on receiving their registrations the core router needs to contact a special access control server residing in the local domain, which decides whether or not to accept the sending requests.

For each domain, a unique border router (*BR*) is elected as the "policy agent" and keeps the entire *SACL* for all the senders in the local domain, and we name this *BR Designated Border Router* (*DBR*) for the domain. In fact the *DBR* can be regarded as core of the sub-tree in the local domain. In this sense, all the data from an upstream domain can only be injected into the local domain from the unique *DBR* and all the senders in this domain can only use this *DBR* to send data up towards the core. This mechanism abides to the "3rd party independence" policy in that data from any sender must be internally delivered to all the local receivers without flowing out of the domain. This requires that joins from different hosts (including both senders and receivers) merge at a common point inside the domain. In *BGP*-4, all the *BRs* of a stub domain know for which unicast prefix(es) each of them is acting as the egress router, this satisfies the above requirement of "path convergence" of internal joins.

Since individual sender authentication is performed within each domain and invalid data never gets any chance to flow out of the local domain, the on-tree *BR* of the upstream domain will always trust its downstream *DBR* and assumes that all the data packets coming from it are originated from authorized senders. Hence, when a packet leaves its local domain and enters remote domains, no further authentication is needed. This also avoids constant lookups when the authenticated data is traveling on the bi-directional tree.

4.2   Inter-domain *SACL* Construction and Activation

Since Border Gateway Multicast Routing (*BGMP [16]*) has been considered as the long-term solution to the Inter-domain multicast routing, in this section we will take *BGMP* as an example to illustrate how sender access control policy can be deployed in inter-domain applications.

First we will discuss how the *DR* for a group member sender submits its join request and how it is added to the *SACL* and activated. This applies to both SR-members and SO-members, the only difference between the two being whether or not to add the interface from which the join-request was received to the interface list with the group state. Only if an on-tree router receives a join request from a sender in the local domain, it will add this sender to its *SACL*, otherwise the router will just forward the join request towards the core without updating its local *SACL*.

In Fig. 3, when host *S* wants to become a SO-member to send data, its *DR* (router *A*) sends a join request towards the *DBR* router *B*, which has the best exit to the root domain. All the internal routers receiving this request will add *S* into their local *SACLs*. Since *B* is the core of the sub-tree for the local domain, it also needs to create a *SACL* entry for host *S* once it receives the join request from its Multicast Interior Gateway Protocol (*M-IGP*) component. Thereafter, *B* finds in its Group Routing Information Base (*G-RIB*) that the best route to the root domain is via its external peer *C* in the transit domain, so router *B* will send the *BGMP* join request towards *C* via its *BGMP* component. Once router *C* receives the join request, it creates (**, G*) state (if it has not been on the tree), but will not create an entry for *S* in its local *SACL*. When *C* finds out that the best exit toward the root domain is *D*, it just forwards the join request to this internal *BGMP* peer, and hence router *D* becomes the *DBR* of the transit domain for group *G*. Suppose *Bidir-PIM* is the *MIGP*, the *RP* in this transit domain should be placed at *D*, and router C will use its *M-IGP* component to send the join request towards *D*. When this join request travels through the transit domain, none of the internal routers along the way in the domain will add *S* into their local *SACLs*. After the join request reaches the root domain and the core router *F* authorizes the new sender by contacting the access control server and sends back the activating-notification, all the on-tree routers (including internal on-tree routers and the *DBR*) in the transit domain just forward it back towards the local domain where the new sender *S* is located. When the packet enters the local domain, all the on-tree routers (namely *B* and *A* in Fig. 3) will activate *S* in their *SACLs*.

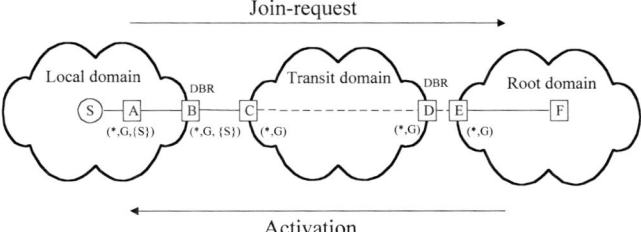

Fig. 3. Inter-domain Join-Request.

As we have also mentioned, a send-only host may also choose to act as a Non-Member Sender (*NMS*). However there are some restrictions when inter-domain multicast routing is involved. If a send-only host is located in the domain where there are no receivers (we call this domain a *send-only domain*), then the host should join the bi-directional tree as a SO-member other than a Non-Member Sender (*NMS*). Otherwise if the host acts as a *NMS*, its registration packet will not hit any on-tree router until it enters remote domains. This forces the on-tree router there to add the sender that is from another domain to its local *SACL*, which does not conform to the rule that on-tree routers only maintain access policy for senders in the local domain. On the other hand, if the host joins as a SO-member and since its *DR* will be on the bi-directional tree, the authentication can be achieved by the on-tree routers in the local domain.

4.3   An Example for Inter-domain Access Policy

An example for inter-domain sender access control is given in Fig. 4. *C* is the core router and domains *X*, *Y* and *Z* are remote domains regarding the core *C*. Hosts *a*, *b*, *c* and *d* are attached to the routers in different domains. Also suppose that host *a* only wants to receive data from the group, hosts *b* and *c* want to both send and receive, while host *d* only wants to send messages to the group without receiving any data from it. In this case, *X* is a receive-only domain and *Z* is a send-only domain. *X1*, *Y1* and *Z1* are border routers that have been selected as the *DBR* for each domain. According to our inter-domain access control scheme, on-tree routers have the *SACL* entry for downstream senders in the local domain, and each *DBR* has the policy for all the senders in the local domain. Hence, *Y1* has the entry for hosts *b* and *c* in its *SACL* while the *SACL* of *X1* contains no entries at all. Although *X* is the parent domain of *Y* and *Z* which both contain active senders, all the on-tree routers in *X* need not add these remote senders to their *SACL*. In fact data coming from *Y* and *Z* has already been authenticated by their own *DBRs* (namely *Y1* and *Z1*) before it flows out of the local domains. Since host *d* only wants to send data to the group and there are no other receivers in domain *Z*, as we have mentioned, host *d* should join as a send-only member. Otherwise if *d* acts as a non-member sender and submits its registration packet towards the core, this makes the first on-tree router (*X2*) add *d* to its *SACL*, however this is not scalable because on-tree routers are forced to add senders from remote domains. On the other hand, if host *d* joins as a send-only member, the shared tree will span to its *DR*, namely *Z2*, and then the authentication can be performed at the routers in the local domain.

As we know, *BGMP* also provides the mechanism for building source-specific branches between border routers. In Fig. 4, we suppose that the current *M-IGP* is *PIM-SM*. At certain time the *DR* in domain *Y* such as *Y3* or *Y4* may wish to receive data from host *d* in domain *Z* via the shortest path tree. Hence (*S, G*) state is originated and passed to the border router *Y5*, which is not the current *DBR* of domain *Y*. When *Y5* receives the source specific join, it will create (*S, G*) state and then send the corresponding *BGMP* source specific join towards *Z1*. On the other hand, since *Z1* is the *DBR* of domain *Z*, intra-domain sender authentication has been performed before the traffic is sent to *Z1's BGMP* component for delivery to remote domains. In fact *Y5* will only receive and accept data originated from host *d* in domain *Z* due to its (*S, G*) state filtering. Once *Y5* receives the data from host *d*, it can directly forward it to all the receivers in the local domain, since *RPF* check can be passed. When the *DR* receives the data from *d* via the shortest path, it will send a source specific prune message up towards the root domain to avoid data duplication. It should be noted that (**, G*) state should only exist in the *DBR* for each domain/group, and internal nodes may only receive source specific traffic via alternative border routers. From this example, it is observed that source specific tree can also interoperate with the proposed sender access control in the receiver's domain (note that the *MIGP* in domain *Y* is not bi-directional routing protocol).

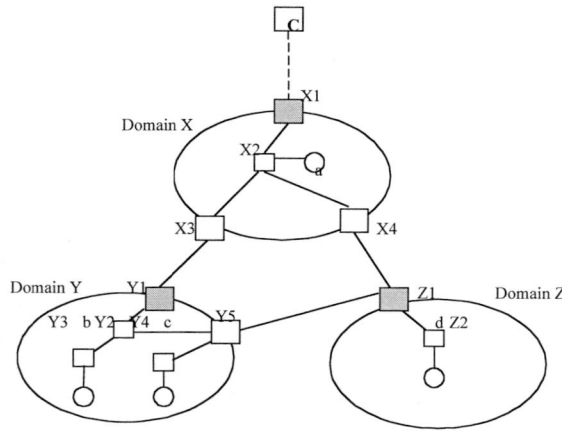

Fig. 4. Example for Inter-domain Sender Access Control.

5   Operations on Multi-access Networks

We need special consideration for protecting group members from unauthorized sources attached to multi-access networks such as *LANs*. As we have mentioned, if an on-tree router receives data packets from its upstream interface, it will always forward them to all the other interfaces with group state, since these packets have been assumed to come from an authorized information source. However this may not be the case if the upstream interface of an on-tree router is attached to a broadcast network. When an unauthorized host wants to send data with group address to the multi-access *LAN*, a corresponding mechanism must be provided to prevent these packets from being delivered to all the downstream group members. To achieve this, once the Designated Router (*DR*) on the *LAN* receives such a packet from its downstream interface, if it cannot find a matching access entry for the data source in its *SACL*, it will discard the packet, and at the same time this *DR* will send a type of "forbidding" control packet containing the unicast address of the unauthorized host to the *LAN* from its downstream interface. Take the *CBT* routing protocol as an example, the IP destination address of this forbidding packet should be "all-cbt-router address (224.0.0.15)" and the value of *TTL* is set to 1. Once the downstream router receives this packet on its upstream interface, it will stop forwarding the data with this unicast address that originates from an unregistered host attached to the *LAN*. Hence all the downstream session members will only receive little amount of useless data for a short period of time. In terms of implementation, the downstream on-tree routers should maintain a "forbidding list" of unauthorized hosts recorded. Since all the possible unauthorized hosts can only come from the local *LAN*, this list will not introduce much overhead to the routers. In Fig. 5, suppose the unauthorized host *S* sends data to the group. When the *DR* (router *A*) cannot find the corresponding entry in its local *SACL*, it immediately discards the packet and then sends a "forbidding" packet containing the address of *S* onto the *LAN*. Once the downstream router *B* receives the forbidding packet, it will stop forwarding data coming from host *S*.

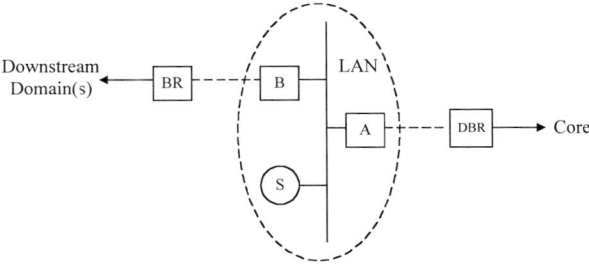

Fig. 5. Access Control Operation on *LANs*.

In inter-domain routing, further consideration is necessary for data traffic traveling towards the core. This is because routers in transit domains do not have *SACL* entry for remote senders in their *SACLs*. Also take Fig. 5 as an example, suppose that the *LAN* is located in a transit domain where there are no local authorized senders, and hence router *A*'s *SACL* is empty. If there is data appearing on the *LAN* destined to the group address, there are only two possibilities: (1) the data came from a downstream domain and was forwarded to the *LAN* by router *B*; (2) a local unregistered host attached to the *LAN* (e.g., host *S*) sent the data. It is obvious that in the former case router *A* should pick up the packet and forward it towards the core, and for the latter, it should just discard the packet and send the corresponding "forbidding" packet onto the *LAN*. Hence this requires that the router be able to distinguish between packets coming from remote domains and packets coming from directly attached hosts on the *LAN*. However, this is easy to achieve by simply checking the source address prefix.

6   *SACL* Scalability Analysis

In this section we discuss scalability issues regarding router memory consumption. For simplicity we only discuss the situation of intra-domain routing here. Nevertheless, when inter-domain hierarchical sender access control is involved, the situation can be improved still further. It is obvious that the *maximum* memory space needed in maintaining a *SACL* is $O(ks)$ where k is the number of multicast groups and s is the number of senders in the group. Typically this is exactly the size of *SACL* in the core router. However, since on-tree routers need not keep the access policy for all sources but only for downstream senders, the average size of *SACL* in each on-tree router is significantly smaller.

We can regard the bi-directional shared tree as a hierarchical structure with the core at the top level, i.e., level 0. Since each of the on-tree routers adds its downstream senders to its local *SACL*, then the *SACL* size S of router i in the shared tree T can be expressed as follows:

$$S(i) = \sum_{(i,j) \in T} (S(j)) \tag{1}$$

and the average *SACL* size per on-tree router is:

$$\overline{S} = \frac{\sum_{i=0}^{H} \sum_{j=1}^{L_i} S(j)}{\sum_{i=1}^{n} Y_i} \tag{2}$$

where H is the number of hops from the farthest on-tree router (or maximum level) and L_i is the number of routers on level i, while

$$Y_i = \begin{cases} 1 & \text{if router } i \text{ is included in the shared tree} \\ 0 & \text{otherwise} \end{cases} \tag{3}$$

To ensure that the scalability issues are fairly evaluated throughout our simulation, random graphs with low average degrees, which represent the topologies of common point-to-point networks, e.g., *NSFNET*, are constructed. Here we adopt the commonly used Waxman's random graph generation algorithm [22] that has been implemented in *GT-ITM*, for constructing our network models. For simplicity, we only consider intra-domain routing scenarios in our simulation.

First we study the relationship between average *SACL* size and total number of senders. In the simulation we generate a random network with 100 routers with the core router also being randomly selected. The number of senders varies from 10 to 50 in steps of 10 while the group size is fixed at 50. Here we study three typical situations regarding the type of sending hosts:

(1) All senders are also receivers (*AM*);
(2) 50% senders are also receivers (*HM*);
(3) None of the senders are receivers (*NM*).
      All send-only hosts choose to act as Non-Member Senders (*NMS*).

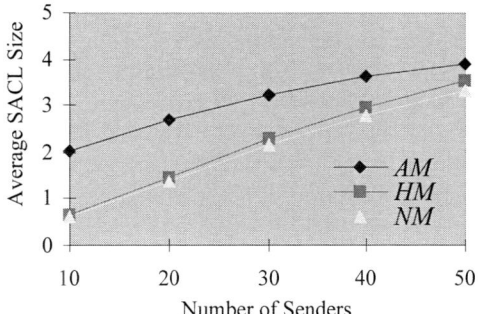

Fig. 6. *SACL* size *vs.* Number of Senders (I).

From Fig. 6 we can see that the average *SACL* size grows as the number of senders increases. However, it can be observed that even when the number of senders reaches a size as large as 50, the average *SACL* size is still very small (less than 4 in size on average). This is in significant contrast with the strategy of "full policy maintenance" (*FPM*) on each router [6, 19]. Further comparison between the two methods is presented in Table 1. From the figure we can also find that if all the senders are also receivers on the bi-directional tree (case *AM*), this results in a larger average *SACL* size. On the other side, if none of the senders is a receiver (case *N M*), the corresponding *SACL* size is smaller. This phenomenon is expected because given the fixed number of receivers on the bi-directional tree as well as the sender group, the larger the proportion of senders coming from receiver set, the larger the resulting average *SACL* size. However this gap decreases with larger sender group size.

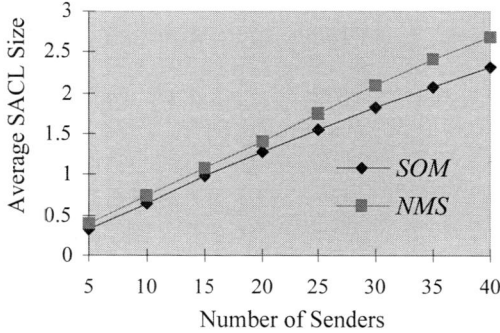

Fig. 7. *SACL* size *vs.* Number of Senders (II).

Next we study the effect on *SACL* size resulting from the senders' choice of acting as a Send-Only Member (*SOM*) or a Non-Member Sender (*NMS*). As we have mentioned, a host only wishing to send data to the group can decide to act as a *SOM* or *NMS*. Fig. 7 illustrates the relationship between the *SACL* size and total number of senders. The group size is fixed at 50 and the number of senders varies from 5 to 40 in steps of 5. It should be noted that in this simulation all group members are receive-only hosts and do not send any data to the group. From the figure we can see that the *SACL* size also grows with the increase of the number of senders. Moreover, if all the hosts join the bi-directional tree and act as Send-Only Members (*SOM*), the average *SACL* size is smaller. The reason for this is obvious: If the hosts choose to take the role of *SOM*, this will make the bi-directional tree expand for including the *DRs* of these senders. Since the number of on-tree routers grows while the total number of senders remains the same, the resulting average *SACL* size will become smaller. On the other hand, if all of the hosts just act as Non-Member-Senders, the figure of the shared tree will not change and no more on-tree routers are involved.

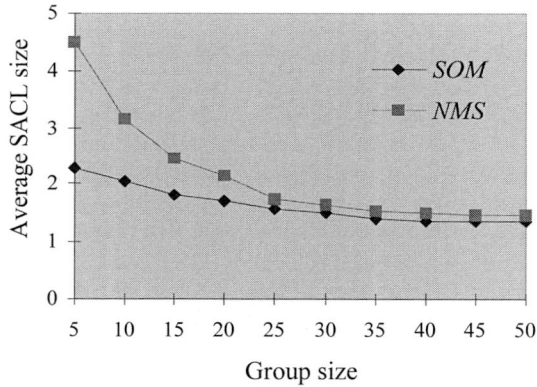

Fig. 8. Average *SACL* Size *vs.* Group Size.

We continue to study the relationship between the average *SACL* size and the group size (number of receivers) with number of senders fixed at 20. We still let these senders choose to act as a *SOM* or *NMS* respectively. From Fig. 8 we can see that the *SACL* size decreases with the growth of the group size in both cases. On the other hand, *SOM* join results in smaller average *SACL* size compared with *NMS*. The gap is more significant when there are fewer receivers. This is because if senders choose to act as *SOM*, they have to join the tree and generate many send-only branches, i.e., more routers are involved in the bi-directional tree. If the hosts just send data without becoming group members, the shared tree won't span to any of these senders, so that the number of on-tree routers is independent of the number of senders. When the group size is small (e.g., 5 receivers), the size of the bi-directional tree will be increased significantly to include all the senders if they join as *SOMs*. This explains why the gap is more obvious when a small set of receivers is involved.

Table 1. Comparison with *FPM*.

S	10	20	30	40
FPM	10	20	30	40
SOM	0.65	1.27	1.82	2.3
NMS	0.73	1.4	2.09	2.73

Finally we give the comparison between our method and the "full policy maintenance" (*FPM*) strategy regarding router's memory consumption. Table 1 gives the relationship of *SACL* size and total number of senders (*S*). From the table we can see that the length of the access list recorded in each on-tree router in *FPM* mechanism is exactly the number of active senders. This imposes very big overhead on routers compared with our proposed scheme. Although the core router also has to maintain the full access list in our method when intra-domain routing is considered, the situation could be improved in large-scale multicast applications by hierarchical control in inter-domain routing which we introduced in section 4.

7   Summary

In this paper we propose an efficient mechanism of sender access control for bi-directional multicast trees in the IP multicast service model. Each on-tree router dynamically maintains access policy for its downstream senders. Under such type of control, data packets from unauthorized hosts are discarded once they hit any on-tree router. In this sense, group members won't receive any irrelevant data, and network service availability is guaranteed since the multicast tree is protected from denial-of-service attacks such as data flooding from any malicious host. In order to achieve scalability for large-scale multicast applications with many information sources and to accommodate more concurrent multicast sessions, we also extend our control mechanism to inter-domain routing where hierarchical access policy is maintained on the bi-directional tree. Simulation results also show that the memory overhead of our scheme is quite light so that good scalability can be achieved.

Nevertheless, this paper only provides a general paradigm of sender access control, but does not present a solution to the restriction of sources based on the specific interest from individual receivers. Related works include [12], [17] and [18], and this will be one of our future research directions.

References

[1]   K. C. Almeroth, "The Evolution of Multicast: From the MBone to Inter-domain Multicast to Internet2 Deployment", IEEE Network special issue on Multicasting, Jan., 2000.

[2]   T. Ballardie, P. Francis, J. Crowcroft, "Core Based Trees (CBT): An Architecture for Scalable Multicast routing", Proc. SIGCOMM'93, pp. 85-95.

[3]   A. Ballardie, "Scalable Multicast Key Distribution", RFC 1949, May 1996.

[4]   A. Ballardie, J. Crowcroft, "Multicast-Specific Security Threats and Counter-measures", Proc. NDSS'95, pp. 2-16 .

[5]   S. Bhattacharyya et al, "An Overview of Source-Specific Multicast (SSM) Deployment", Internet Draft, draft-ietf-ssm-overview-*.txt, May 2001, work in progress.

[6]   B. Cain, "Source Access Control for Bidirectional trees", 43rd IETF meeting, December, 1998.

[7]   B. Cain et al, "Internet Group Management Protocol, Version 3", Internet draft, draft-ietf-idmr-igmp-v3-*.txt, Feb. 1999, work in progress.

[8]   S. Deering et al, "The PIM Architecture for Wide-Area Multicast Routing", IEEE/ACM Transactions on Networking, Vol. 4, No. 2, Apr. 1996, pp. 153-162.

[9]   S. Deering, "Multicast Routing in Internetworks and Extended LANs", Proc. ACM SIGCOMM, 1988, pp. 55-64.

[10] C. Diot et al, "Deployment Issues for the IP Multicast Service and Architecture", IEEE Network, Jan./Feb. 2000, pp 78-88.

[11] W. Fenner, "Internet Group management Protocol, version 2", RFC 2236, Nov. 1997.

[12] B. Fenner et al, "Multicast Source Notification of Interest Protocol (MSNIP)", Internet Draft, draft-ietf-idmr-msnip-*.txt, Feb. 2001.

[13] M. Handley et al, "Bi-directional Protocol Independent Multicast (BIDIR-PIM)", Internet Draft, draft-ietf-pim-bidir-*.txt, Nov. 2000, work in progress.

[14] H. W. Holbrook, D. R. Cheriton, "IP Multicast Channels: EXPRESS Support for Large-scale Single-source Applications", Proc. ACM SIGCOMM'99.

[15] H. W. Holbrook, B. Cain, "Source-Specific Multicast for IP", Internet Draft, draft-holbrook-ssm-arch-*.txt, Mar. 2001, work in progress.

[16] S. Kummar et al, "The MASC/BGMP Architecture for Inter-domain Multicast Routing", Proc. ACM SIGCOMM'99.

[17] B. N. Levine et al, "Consideration of Receiver Interest for IP Multicast Delivery", Proc. IEEE INFOCOM 2000, vol. 2, pp. 470-479.

[18] M. Oliveira et al, "Router Level Filtering for Receiver Interest Delivery", Proc. NGC' 2000.

[19] R. Perlman et al, "Simple Multicast: A Design for Simple, Low-overhead Multicast" Internet Draft, draft-perlman-simple-mulitcast-*.txt, Oct. 1999, work in progress.

[20] C. Rigney et al, "Remote Authentication Dial In User Service (RADIUS)", RFC 2138, Apr. 1997.

[21] C. Shields et al, "KHIP-A Scalable Protocol for Secure Multicast Routing", Proc. ACM SIGCOMM'99.

[22] B.M. Waxman, "Routing of multipoint connections", IEEE JSAC 6(9) 1988, pp. 1617-1622.

EHBT: An Efficient Protocol for Group Key Management*

Sandro Rafaeli, Laurent Mathy, and David Hutchison

Computing Department, Lancaster University, LA1 4YR, Lancaster, UK

Abstract. Several protocols have been proposed to deal with the group key management problem. The most promising are those based on hierarchical binary trees. A hierarchical binary tree of keys reduces the size of the rekey messages, reducing also the storage and processing requirements. In this paper, we describe a new efficient hierarchical binary tree (EHBT) protocol. Using EHBT, a group manager can use keys already in the tree to derive new keys. Using previously known keys saves information to be transmitted to members when a membership change occurs and new keys have to be created or updated. EHBT can achieve $(I \cdot \log_2 n)$ message size (I is the size of a key index) for join operations and $(K \cdot \log_2 n)$ message size (K is the size of a key) for leave operations. We also show that the EHBT protocol does not increase the storage and processing requirements when compared to other HBT schemes.

1 Introduction

With IP multicast communication, a group message is transmitted to all members of the group. Efficiency is clearly achieved as only one transmission is needed to reach all members. The problems start because any machine can join a multicast group and start receiving the messages sent to the group without the sender's knowledge. This characteristic raises concerns about privacy and security since not every sender wants to allow everyone to have access to its communication.

Cryptographic tools can be used to protect group communication. An encryption algorithm takes input data (e.g. a group message) and performs some transformations on it using a key (where the key is a randomly generated number). This process generates a ciphered message. There is no easy way to recover the original message from the ciphered text other than by knowing the key [9].

When applying such technique, it is possible to run secure multicast sessions. Group messages are protected by encryption using a chosen key (*group key*). Only those who know the group key are able to recover the original message. However, distributing the group key to valid members is a complex problem. Although rekeying a group before the join of a new member is trivial (send the new group key to the old group members encrypted with the old group key),

---

* The work presented here was done within the context of ShopAware - a research project funded by the European Union in the Framework V IST Programme.

J. Crowcroft and M. Hofmann (Eds.): NGC 2001, LNCS 2233, pp. 159–171, 2001.

rekeying the group after a member leaves is far more complicated. The old key cannot be used to distribute a new one, because the leaving member knows the old key. A group manager must, therefore, provide other scalable mechanisms to rekey the group.

Several researchers have studied the use of a hierarchical binary tree (HBT) for the group key management problem. Using an HBT, the key distribution centre (KDC) maintains a tree of keys, where the internal nodes of the tree hold key encryption keys (KEKs) and the leaves correspond to group members. Each leaf holds a KEK associated to that one member. Each member receives and maintains a copy of the KEK associated to its leaf and the KEKs correspondent to each ancestor node in the path from its parent node to the root. All group members share key held by the root of the tree. For a balanced tree, each member stores $\log_2 n + 1$ keys, where n is the number of members. This hierarchy is explored to achieve better performance when updating keys.

In this paper, we propose a protocol to efficiently built an HBT, which we call the EHBT protocol. The EHBT protocol achieves $(I \cdot \log_2 n)$ message size for addition operations and $(K \cdot \log_2 n)$ message size for removal operations keeping the storage and processing on both, client and server sides to a minimum. We achieve these bounds using well-known techniques, such as a one–way function and the *xor* operator.

2   Related Work

Wallner et al [13] were the first to propose the use of an HBT. In their approach, every time the group membership changes, internal node keys (affected by the membership change) are updated and every new key is encrypted with each of its children's keys and then multicast. A rekey message conveys $2 \cdot \log_2 n$ keys for including or removing a member.

Caronni et al [12] proposed a very similar protocol to that of Wallner, but they achieve a better performance regarding the size of multicast messages for joining operations. We refer to this protocol as HBT+. Instead of encrypting new key values with their respective children's key, Caronni proposes to pass those keys into a one–way function. Only the indexes of the refreshed keys need to be multicast and an index size is smaller than the key size.

An improvement to the hierarchical binary tree approach is the one–way function tree (OFT) proposed by McGrew and Sherman [5]. The keys of a node's children are *blinded* using a one–way function and then mixed together using the xor operator. The result of this mixing is the KEK held by the node. The improvement is due to the fact that when the key of a node changes, its blinded version is only encrypted with the key of its sibling node. Thus, the rekey message carries just $\log_2 n$ keys.

Canetti et al [3] proposed a slightly different approach that achieves the same communication overhead. Their scheme uses a pseudo–random–generator (PRG) [9] to generate the new KEKs rather than a one–way function and it is applied only on user removal.

Perrig et al proposed the efficient large–group key (ELK) protocol [6]. The ELK protocol is very similar to the OFT, but ELK uses pseudo–random functions (PRFs)[1] to build and manipulate the keys in the tree. ELK employs a timely rekey, hence, at every time interval, the KDC refreshes the root key using the PRF function and then uses it to update the whole key tree. By deriving all keys, ELK does not require any multicast messages during a join operation. All members can refresh their own keys, hence no rekey message is required. When members are deleted, as in OFT, new keys are generated from both its children's keys.

3   Efficient Hierarchical Binary Tree Protocol

In the EHBT protocol, a KDC maintains a tree of keys. The internal nodes of the tree hold KEKs and the leaves correspond to group members. Keys are indexed by randomly chosen numbers. Each leaf holds a secret key that is associated to that member. The root of the tree holds a common key to all members.

Ancestors of a node are those nodes in the path from its parent node to the root. The set of ancestor of a node is called *ancestor set*. Each member knows only its own key (associated to its leaf node) and keys correspondent to each node in its *ancestor set*. For a balanced tree, each member stores $\log_2 n + 1$ keys, where n is the number of members.

In order to guarantee backward and forward secrecy [11], the keys related to joining members or leaving members should be changed every time the group membership changes. The new keys in the *ancestor set* of an affected leaf are generated upwards from the key held by the affected leaf's sibling up to the root. Using keys that are already in the tree can save information to be transmitted to members when a membership occurs and new keys have to be created or updated.

The formula $\mathcal{F}(x, y) = h(x \oplus y)$ is used to generate keys from other keys, where h is a one–way hash function and $\oplus$ is a normal xor operator. The obvious functionality of function h is to *hide* the original value of x and y into value z in a way that if one knows only z he cannot find the original values x and y. The functionality of $\oplus$ is to mix x and y and generate a new value.

We say that a key k_i can be *refreshed* by doing $k_i' = \mathcal{F}(k_i, i)$, where i is the index (or identifier) of key k_i or key k_i can be *updated* by deriving one of its children key by doing $k_i' = \mathcal{F}(k_i^{left|right}, i)$, where $k_i^{left|right}$ is the key of i's either left or right child. Appendix A describes the reason for using index i in function $\mathcal{F}$.

3.1   Rekey Message Format

A member can receive two types of information in a rekey message, one telling him to refresh or update the value of a key, the other telling him the new value

---

[1] ELK uses the stream cipher RC5 [8] as the PRF.

of a key. In the former case, the member receives an *id* and in the latter case, he receives a key value. After deriving a key, a member will try to derive all other keys by himself (from that key up to the root) unless he receives another information telling him something different. For example, if key K_i is refreshed, the KDC needs to send to K's holders the identification of the key so that they can perform the refresh operation themselves. Or, if a node n has its key updated $(K'_n = \mathcal{F}(k_L, n))$, then it implies sending to member L the index n and to the other child, namely R, the new key value K'_n (because R does not know L's key).

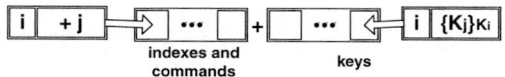

Fig. 1. Example of a Rekey Message.

The rekey message that relays this information has two parts. The first part carries commands and the second carries keys. Each piece of information is indexed by a key index. Keys are encrypted with the key indicated by the key index (see Figure 1), but commands are not encrypted because they do not carry vital information. Based on commands and keys, members can find out which keys they must refresh or update, or just substitute, because they have received a new key value to a specific key.

Algorithm 1: Reading Rekey Message Algorithm.
(1)   receive rekey message
(2)   set last command to "keep key"
(3)   **while** there is a key to be derived
(4)   get a key index from key–list
(5)   search indexes part of rekey message for key index
(6)   **if** there is a command
(7)   execute the command on the specific key
(8)   set last command to this command
(9)   **else**
(10)   search keys part of the rekey message for key index
(11)   **if** there is a key
(12)   substitute it in the key list
(13)   set last command to "update"
(14)   **if** there is no command or key
(15)   execute last command in current key

The algorithm to handle rekey messages starts with a member holding a list of known keys (key–list). After executing the algorithm, a member will have all his keys freshened up. A simplified version of this algorithm appears in Algorithm 1.
    In the remainder of this paper, we use the following notation:

$+i$ or $-i$ or $*i$	are commands to be applied on key i
$\mathcal{R}(k_i)$	refresh k_i applying $\mathcal{F}(k_i, i)$
$\mathcal{U}(k_i)$	update k_j applying $\mathcal{F}(k_i, j)$
$\{x\}_k$	encryption of x with k
$j : command$	command to key j's holder
$[commands, keys]$	message containing commands and keys

4   Basic Operations

In this section, we describe the basic algorithms for join and leave operations for single and multiple cases.

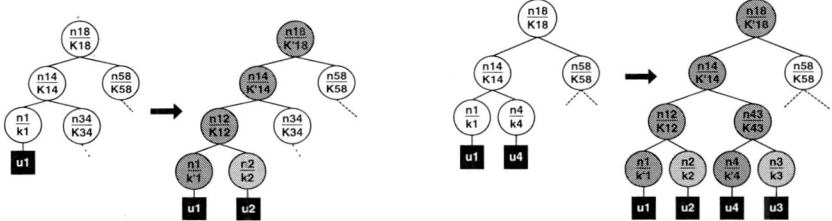

Fig. 2. User u_2 Joins the Tree.   **Fig. 3.** Users u_2 and u_3 Join the Tree.

Single Member Join Algorithm. When a member joins the group, it is associated to a leaf node n. The KDC assigns a randomly chosen key k_n to n. Leaf n is then included in the tree at the parent of the shallowest leaf node s (to keep the tree as short as possible). Leaf s is removed from the tree, and in its place a new node p is inserted. Leaves s and n are inserted as p's children. We see an example in Figure 2: Member 2 is placed in leaf n_2, which is inserted at node n_{12}. Node n_{12} becomes the new parent of leaves n_1 and n_2. Leaf n_2 is assigned key k_2.

In order to keep the backward secrecy, keys in n_1's *ancestor set* need to receive new values. Key k_1 is refreshed ($k'_1 = \mathcal{R}(k_1)$), K_{12} receives a value based on k'_1 ($K_{12} = \mathcal{U}(k'_1)$) and keys K_{14} and K_{18} are refreshed ($K'_{14} = \mathcal{R}(K_{14})$ and $K'_{18} = \mathcal{R}(K_{18})$).

Note that during a join operation, keys, which were already in the tree, are just refreshed. Members holding those keys only need to be told those keys' indexes to be able to generate their new values, which means that these keys do not have to be transmitted. In the same way, members that had their keys used for generating new keys just have to be told the index of the new key and they can generate that key by themselves.

The KDC generates unicast messages for member n_2 ($[k_2, K_{12}, K'_{14}, K'_{18}]$) and member n_1 ($[+12]$), and multicast message $[14 : *14, 18 : *18]$.

Member u_2 receives its unicast message and creates its key–list. Member u_1 receives its unicast message and derives key K_{12}, including it in its key–list. Members holding keys K_{14} and K_{18} refresh these keys.

Multiple Members Join Algorithm. Several new members are inserted in the tree as in the single member join algorithm. They are associated to nodes and the nodes are placed at the parent of the shallowest leaves. However, the keys in the tree are modified in a slightly different manner. New nodes' *ancestor sets* converge at some point and all keys that are in more than one *ancestor set* are modified only once.

See Figure 3 for an example. Members u_2 and u_3 joined the group and have been placed at nodes n_{12} and n_{43}, respectively. Following the single member join algorithm, the keys in member u_2's *ancestor set* are changed: first, $k_1' = \mathcal{R}(k_1)$, and then, $K_{12} = \mathcal{U}(k_1')$, $K_{14} = \mathcal{R}(K_{14})$, $K_{18}' = \mathcal{R}(K_{18})$. In the same way, keys in member u_3's *ancestor set* are changed: first, $k_4' = \mathcal{R}(k_4)$, and then, $K_{43} = \mathcal{U}(k_4')$. Keys K_{12} and K_{18} have already been changed because of member u_2, hence they are not changed again.

The KDC generates unicast messages for member n_2 ($[k_2, K_{12}, K_{14}', K_{18}']$), member n_3 ($[k_3, K_{43}, K_{14}', K_{18}']$), member u_1 ($[+12]$) and member u_4 ($[+43]$), and multicast message $[14 : *14, 18 : *18]$.

Members u_2 and u_3 receive their unicast messages and create their respective key–lists. Member u_1 receives the unicast message, derives key K_{12}, and includes it in its key–list. Member u_4 does the same with key K_{43}. Members holding keys K_{14} and K_{18} refresh these keys.

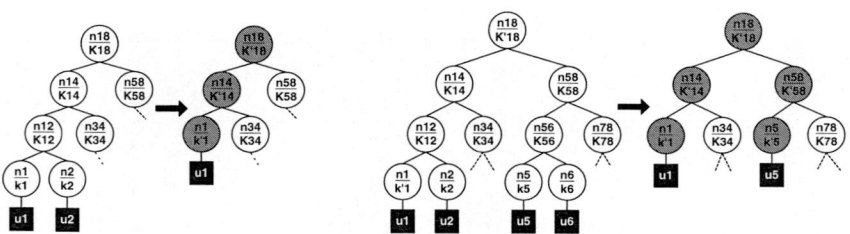

Fig. 4. User u_2 Leaves the Tree.   **Fig. 5.** Users u_2 and u_6 Leave the Tree.

Single Member Leave Algorithm. When a member u leaves or is removed from the group, its sibling s replaces its parent p. Moreover, all keys known by u should be updated to guarantee forward secrecy. For example, see Figure 4: u_2 leaves (or is removed from) the group and its node is removed from the tree. Node n_{12} is also removed and leaf n_1 is promoted to its place.

In order to keep the forward secrecy, keys in n_1's *ancestor set* need to receive new values. Keys K_{14} and K_{18} have to be updated: $k_1' = \mathcal{R}(k_1)$, $K_{14}' = \mathcal{U}(k_1')$ and $K_{18}' = \mathcal{U}(K_{14}')$.

Note that in removal operations, all keys in the removed member's *ancestor set* are updated. Those keys cannot be just refreshed because the removed member knows their previous values and could easily calculate the new values. Since the new values are all generated from the removed member's sibling key, which was not known by the removed member, the removed member cannot find the new values.

The KDC generates multicast message $[1 : -12, \{K_{14}'\}_{K_{34}}, \{K_{18}'\}_{K_{58}}]$.

Member n_1 refreshes k_1' and, because it has removed K_{12}, it updates K_{14} and K_{18}. Members holding key K_{34} get new key K_{14}' and then update key K_{18}. Members holding key K_{58} get new key K_{18}'.

Multiple Members Leave Algorithm. This algorithm is handled similarly to the single member leave algorithm. The leaving nodes are removed and the tree shape is adjusted accordingly. As in the multiple join algorithm, there can be several different path from removed nodes to the root, which means that the root key can be updated by several nodes (see Figure 5).

In order to avoid several root key versions for the same operation, the KDC chooses one of the paths and use it to update the root key. For example, in Figure 5, n_2 and n_6 leave the group and nodes n_1 and n_5 are promoted to their respective parents' places (n_{12} and n_{56}). Both are used to derive their new parent keys K_{14}' and K_{58}', but then they both cannot be used to update key K_{18}'. In this case, the KDC chooses one of them to update key K_{18}' and the other will receive the updated key. For instance, the KDC chooses node n_1 and then the keys are updated as follows: $k_1' = \mathcal{R}(k_1)$, $K_{14}' = \mathcal{U}(k_1')$, $K_{18}' = \mathcal{U}(K_{14}')$, $k_5' = \mathcal{R}(k_5)$ and $K_{58}' = \mathcal{U}(k_5')$.

The KDC generates multicast message $[1 : -12, 5 : -56, \{K_{14}'\}_{K_{34}}, \{K_{58}'\}_{K_{78}}, \{K_{18}'\}_{K_{58}'}]$.

Member n_1 refreshes k_1' and, because it has removed K_{12}, it updates K_{14}' and K_{18}'. Key K_{34}'s holders recover K_{14}' and update K_{18}'. Member n_5 refreshes k_5' and updates K_{58}', but since there is a new key encrypted with K_{58}', n_5 stops updating its keys and just recovers K_{18}'. Key K_{78}'s holders recover K_{58}' and, since there is a key encrypted with it, they just recover K_{18}'.

Rebalancing. The efficiency of the key tree depends crucially on whether the tree remains balanced or not. A tree is said to be balanced if no leaf is much further away from the root than any other leaf. In general, for a balanced binary tree with n leaves, the distance from the root to any leaf is $\log_2 n$, but if the tree is unbalanced, the distance from the root to a leaf can become as high as n. Therefore, it is desirable to keep a key tree as balanced as possible.

The rebalancing works by getting the shallowest and deepest internal nodes and comparing their depths. If the depth gap is larger than two then it means that the tree is unbalanced and needs to be levelled. For balancing the tree, the deepest leaf node is removed, which makes its sibling to go one level up (similarly to the removing algorithm), and inserted at the shallowest node (similarly to the inserting algorithm). This procedure is repeated until the difference between the depths of the shallowest and the deepest nodes is smaller than two.

In a rebalancing operation, the deepest node, which has been moved from one position in the tree to another, requires that its old keys need to be updated (as in a deletion operation) and it needs to have access to the keys in its new path to the root (as in an insertion operation). Therefore, an insertion and a deletion are performed simultaneously.

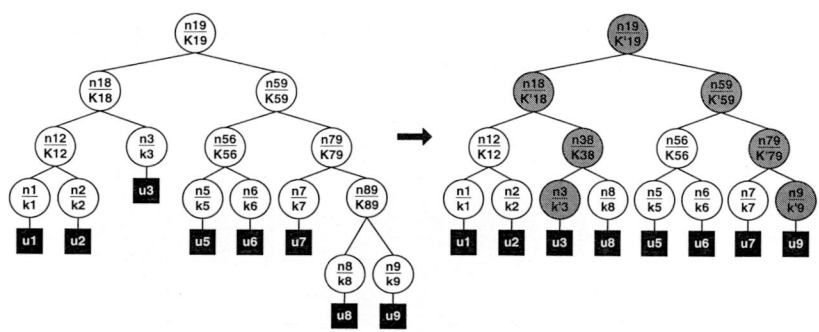

Fig. 6. Rebalancing the Tree.

See Figure 6 for an example. The tree needs a rebalancing, so leaf n_8 is deleted from its original position (n_{89}) and inserted into a new position (n_{38}). The deletion starts a removal operation with leaf n_9 updating the new keys. At the same time, leaf n_3 starts refreshing the keys on its path (as an insertion requires). The new keys are calculated as follows: $k_9' = \mathcal{R}(k_9)$, $K_{79}' = \mathcal{U}(k_9')$, $K_{59}' = \mathcal{U}(K_{79}')$, $k_3' = \mathcal{R}(k_3)$, $K_{38} = \mathcal{U}(k_3')$ and $K_{18}' = \mathcal{R}(K_{18}')$. Key K_{19}' does not need to be changed.

The KDC generates unicast messages for member n_8 ($[K_{38}, K_{18}']$) and member u_3 ($[+38]$), and multicast message $[9 : -89, 18 : *18, \{K_{79}'\}_{k_7}, \{K_{59}'\}_{K_{56}}]$.

Member n_8 deletes all its known keys and replaces them by those just received. Member n_9 updates its keys. Members n_7 and key k_{56}'s holders extract their parts and update their keys. Member n_3 derives K_{38}'. Key K_{18}'s holders refresh K_{18}'.

5   Evaluation

In this section, we compare the properties of the EHBT algorithm with the other algorithms introduced in section 2: PRGT[2] (Canetti et al.), HBT+ (Caronni et

---

[2] Canetti does not specify the PRG function to use, hence we assume the same RC5 algorithm used in ELK.

al), OFT (McGrew and Sherman) and ELK (Perrig). We focus our criteria on KDC computation, joined member computation (for insertions), sibling computation (sibling to the joining/leaving member), size of the multicast message, size of the unicast messages and storage at both KDC and members.

The notations used in this section are:

n	number of member in the group
d	height of the tree (for a balanced tree $d = \log_2 n$)
I	size of a key index in bits
K	size of a key in bits
G	key generation
H	hash function execution
X	xor operation
E	encryption operation
D	decryption operation

Table 1 summarizes the computation required from the KDC, joined member and sibling to joined member, and message size of joining member's unicast message, sibling's unicast message and multicast message during single join operations.

Table 1. Single Join Operation Equations.

Scheme/	Computation			Message size		
Resource	KDC	Join member	Sib member	Join unicast	Sib unicast	Multicast
EHBT	$G + (d+1)(X + H + E)$	$(d+1)D$	$(d+1)(X + H)$	$(d+1)K$	I	dI
PRGT	$2G + dH + (d+1)E$	$(d+1)D$	$D + dH$	$(d+1)K$	$I + K$	dI
HBT+	$2G + dH + (d+1)E$	$(d+1)D$	$D + dH$	$(d+1)K$	$I + K$	dI
OFT	$G + (d+1)H + dX + 3dE$	$(d+1)D + d(H + X)$	$2D + d(H + X)$	$(d+1)K$	$I + 2K$	$(d+1)K$
ELK	$G + (4n-2)E$ and $(d+3)E$	$(d+1)D$	$2dE$ and $2E$	$(d+1)K$	I	0

Table 2 summarises multiple join operation equations. The parameters analysed are the same parameters used in Table 1. The equations are valid for multiple joins when the original number of members is doubled after the mass join, which means that every old member gets a new sibling (a new member) and all the keys in the tree are affected. This represents the worst case possible for join operations. For the sake of the equations in this table, n is the original number of members in the group previously to the mass join, but d is the new height of the tree after the mass join.

EHBT requires less computation than the other schemes, but it loses out to ELK when comparing the message sizes. The reason for that is that ELK employs a timed rekey, which means that the tree is completely refreshed at intervals, despite membership changes, thus only the index of the new parent inserted needs to be sent to the sibling of the joining member. However, this rises two issues: first, at every interval the KDC has to refresh all its 2n-1 keys, which implies unnecessary work for the KDC; second, this scheme does not support rekey on membership changes (regarding join operations). Additionally, ELK imposes some delay on the joining member before he receives the group key.

Table 2. Multiple Join Operation Equations.

Scheme/	Computation			Message size		
Resource	KDC	Join member	Sib member	Join unicast	Sib unicast	Multicast
EHBT	$nG + (3n-1)(X+H) + n(d+1)E$	$(d+1)D$	$(d+1)(X+H)$	$n : (d+1)K$	$n : I$	$(n-1)I$
PRGT	$2nG + (n-1)H + n(d+2)E$	$(d+1)D$	$D + dH$	$n : (d+1)K$	$n : I + K$	$(n-1)I$
HBT+	$2nG + (n-1)H + n(d+2)E$	$(d+1)D$	$D + dH$	$n : (d+1)K$	$n : I + K$	$(n-1)I$
OFT	$nG + (4n-2)(H+X)+$ $(nd+5n-1)E$	$(d+1)D+$ $d(H+X)$	$2D+$ $d(H+X)$	$n : (d+1)K$	$n : I + 2K$	$(2n-2)K$
ELK	$(8n-2)E$ and $nG + n(d+3)E$	$(d+1)D$	$2dE$ and $2E$	$n : (d+1)K$	$n : I$	0

Table 3. Single Leave Operation Equations.

Scheme/	Computation		Multicast
Resource	KDC	Sib member	
EHBT	$d(X+H+E)$	$d(X+H)$	$I + dK$
PRGT	$(2d+1)E$	$D + dE$	$I + (d+1)K$
HBT+	$2dE$	dD	$I + 2dK$
OFT	$d(H+X+E)$	$D + d(H+X)$	$I + (d+1)K$
ELK	$8dE$	$dD + 5dE$	$I + d(n_1 + n_2)$

Table 3 summarizes the KDC computation, sibling computation and multicast message size during single leave operations. We also analyse the equations of multiple leave operations, and we show the results in Table 4. For mass leaving, we consider the situation when exactly half of the group members leave the group. The sibling of every leaving member remains in the tree, and hence, all keys in the tree are affected.

Table 4. Multiple Leave Operation Equations.

Scheme/	Computation		Multicast
Resource	KDC	Sib member	
EHBT	$(2n-1)(X+H) + (n-1)E$	$D + (d+1)(X+H)$	$nI + (n-1)K$
PRGT	$(5n/2 - 2)E$	$D + dE$	$(3n/2 - 1)K$
HBT+	$(2n-2)E$	dD	$nI + 2(n-1)K$
OFT	$(2n-2)H + (n-1)X + (3n-2)E$	$(d+1)D + d(H+X)$	$nI + (3n-2)K$
ELK	$(7n-3)E$	$dD + 5dE$	$nI + (n-1)(n_1 + n_2)$

For leaving operations, again EHBT achieves better results than the other schemes regarding the computations involved, but loses out to ELK when comparing the multicast message size. ELK has a slightly smaller multicast message than EHBT, because it sacrifices security. ELK uses only $n_1 + n_2$ bits of a total K possible bits for generating a new key and this procedure weakens that key, Consequently, an expelled member needs to compute only $2^{n_1+n_2}$ possibilities to recover the new key. In EHBT, however, an expelled member needs to compute the full 2^K operations to brute-force the new key.

We have simulated a group with 8192 members. For the calculations of the multiple join operations, we doubled the size of the group to 16384 members, and then we removed all joining members and finished with the 8192 original members. We measured encryption and decryption times for the RC5 algorithm, MD5 hash function and *xor* operation. We used 16-bit keys for the calculations. We used Java version 1.3 and IAIK [4] cryptographic toolkit on a 850Mhz Mobile Pentium III processor. It takes $1.72 \cdot 10^{-2}$ ms for RC5 to encrypt a 16-bit key with a 16-bit key, and $1.73 \cdot 10^{-2}$ ms to decrypt it. Hashing a 16-bit key takes $4.95 \cdot 10^{-3}$ ms and *xoring* it takes $1.59 \cdot 10^{-3}$ ms. Finally, generating a 16-bit keys takes $7.33 \cdot 10^{-3}$. Applying these numbers into Tables 2 and 4 produces the results in Table 5 that show that EHBT in general is faster to compute than the other protocols.

Table 5. Time in Milliseconds for Multiple Joins and Leaves.

Scheme/	Multiple Join			Multiple Leave	
Resource	KDC	Join member	Sib Member	KDC	Sib member
EHBT	2334	0.25	0.09	248.03	0.10
PRGT	2415	0.25	0.08	352.22	0.24
HBT+	2415	0.25	0.08	281.77	0.22
OFT	2951	0.35	0.12	516.78	0.32
ELK	1140 + 2455	0.25	0.48 + 0.03	1105.46	1.34

Finally, EHBT and the other schemes require the KDC to store $2n - 1$ keys and members to store $d + 1$ keys.

6   Security Considerations

The security of the EHBT protocol relies on the cryptographic properties of the h function. One–way hash functions, unfortunately, are not proven secure [2]; nevertheless, for the time being, there has not been any successful attack on either the full MD5 [7] or SHA [1] algorithms [10].

Taking into account the use of hash functions as function h, attacks on the hidden key are limited to brute-force attack. Such an attack can take 2^n hashes to find the original key, with n being the number of bits of the original key used as input.

In order to guarantee backward secrecy and forward secrecy, every time there is a membership change, the keys related to joining members or leaving members are changed.

When a member is added to the tree, all keys held by nodes in its *ancestor set* are changed to avoid giving the new member access to past information. For example, see Figure 2, when member n_2 is inserted in the tree, key K_{12} is created and keys K'_{14} and K'_{18} are refreshed. Node n_2 does not have access to the old values, because it only receives the new key values, which were hidden by the hash function, and assuming the hash function is secure, n_2 has no other way to recover the old key but brute-forcing it. The same rule applies when n_2 leaves; key K_{12} is deleted from the tree and keys K'_{14} and K'_{18} are updated and since n_2 does not have access to their new values it does no longer has access to the group communication.

7   Conclusion

Using one–way hash functions and *xor* operations, we constructed an efficient HBT protocol that achieves better overall performance than other HBT protocols. Our protocol, called EHBT, requires only $(I \cdot \log_2 n)$ message size for join operations and $(K \cdot \log_2 n)$ message size for leaving operations. Additionally, EHBT requires the same key storage as other HBT protocols, and it requires much less computation to rekey the tree after membership changes.

References

1. N.F.P. 180-1. Secure Hash Standard. National Institute of Standards and Technology, U.S. Department of Commerce, DRAFT, May 1994.
2. S. Bakhtiari, R. Safavi-Naini, and J. Pieprzyk. Cryptographic Hash Functions: A Survey. Technical Report 95-09, University of Wollongong, July 1995.
3. R. Canetti, J. Garay, G. Itkis, D. Micciancio, M. Naorr, and B. Pinkas. Multicast Security: A Taxonomy and Some Efficient Constructions. In *Proc. of INFOCOM'99*, volume 2, pages 708–716, New Yok, NY, USA, March 1999.
4. I.-J. Group. IAIK, java–crypto toolkit. Web site at http://jcewww.iaik.tu-graz.ac.at/index.htm.
5. D.A. McGrew and A.T. Sherman. Key Establishment in Large Dynamic Groups Using One-Way Function Trees. Technical Report No. 0755, TIS Labs at Network Associates, Inc., Glenwood, MD, May 1998.
6. A. Perrig, D. Song, and J. D. Tygar. ELK, a New Protocol for Efficient Large-Group Key Distribution. In *2001 IEEE Symposium on Security and Privacy*, Oakland, CA, USA, May 2001.
7. R. Rivest. The MD5 Message-Digest Algorithm. RFC 1321, April 1992.
8. R. Rivest. The RC5 encryption algorithm. In *Fast Software Encryption, 2nd Int. Workshop*, LNCS 1008, pages 86–96. Springer-Verslag, December 1995.
9. B. Schneier. *Applied Cryptography Second Edition: Protocols, algorithms, and source code in C*. John Wiley & Sons, Inc., 1996. ISBN 0-471-11709-9.
10. W. Stallings. *Cryptography and Network Security*. Prentice–Hall, 1998. ISBN 0-138-69017-0.
11. M. Steiner, G. Taudik, and M. Waidner. Cliques: A new approach to group key agreement. Technical Report RZ 2984, IBM Research, December 1997.
12. M. Waldvogel, G. Caronni, D. Sun, N. Weiler, and B. Plattner. The VersaKey Framework: Versatile Group Key Management. *IEEE Journal on Selected Areas in Communications (Special Issue on Middleware)*, 17(9):1614–1631, August 1999.
13. D. Wallner, E. Harder, and R. Agee. Key Management for Multicast: Issues and Architectures. RFC 2627, June 1999.

A   Reasoning on Using Index i in Function $\mathcal{F}$

Index i is included in the formula $\mathcal{F}$ to avoid giving the possibility for members to have access to keys that they are not meant to. For example, removing member n_2 in Figure 4, means new keys $k_1' = \mathcal{U}(k_1)$, $k_{14}' = \mathcal{R}(k_1')$ and $k_{18}' = \mathcal{R}(k_{14}')$.

If, immediately after member n_2 has left the group, member n_0 joins it and is inserted as a sibling of n_1, then it means new keys $k_1'' = \mathcal{U}(k_1')$, $k_{10} = \mathcal{R}(k_1'')$ (new node n_{10}), $k_{14}'' = \mathcal{U}(k_{14}')$ and $k_{18}'' = \mathcal{U}(k_{18}')$.

If we remove i from function $\mathcal{F}$ and instead only apply a simple hash h to update keys then the keys from the removal above become $k_1' = h(k_1)$, $k_{14}' = h(k_1')$ (or $h(h(k_1))$) and $k_{18}' = h(k_{14}')$ (or $h(h(h(k_1)))$) and the keys from the join become $k_1'' = h(k_1')$ (or $h(h(k_1))$), $k_{10} = h(k_1'')$ (or $h(h(h(k_1)))$), $k_{14}'' = h(k_{14}')$ and $k_{18}'' = h(k_{18}')$. As one can see, key k_{10} and k_{18}' are identical, which means that member n_0 can have access to past messages encrypted with k_{18}' (or k_{10}).

Aggregated Multicast with Inter-Group Tree Sharing[*]

Aiguo Fei[1], Junhong Cui[1], Mario Gerla[1], and Michalis Faloutsos[2]

[1] Computer Science Department, University of California, Los Angeles, CA90095
[2] Computer Science and Engineering, University of California, Riverside, CA 92521

Abstract. IP multicast suffers from scalability problems for large numbers of multicast groups, since each router keeps forwarding state proportional to the number of multicast tree passing through it. In this paper, we present and evaluate *aggregated multicast*, an approach to reduce multicast state. In aggregated multicast, multiple groups are forced to share a single delivery tree. At the expense of some bandwidth wastage, this approach can reduce multicast state and tree management overhead at transit routers. It may also simplify and facilitate the provisioning of QoS guarantee for multicast in future aggregated-flow-based QoS networks. We formulate the tree sharing problem and propose a simple intuitive algorithm. We study this algorithm and evaluate the trade-off of aggregation vs. bandwidth overhead using simulations. Simulation results show that significant aggregation is achieved while at the same time bandwidth overhead can be reasonably controlled.

1 Introduction

Multicast is a mechanism to efficiently support multi-point communications. IP multicast utilizes a tree delivery structure on which data packets are duplicated only at fork nodes and are forwarded only once over each link. Thus IP multicast is resource-efficient in delivering data to a group of members simultaneously and can scale well to support very large multicast groups. However, even after approximately twenty years of multicast research and engineering effort, IP multicast is still far from being as common-place as the Internet itself.

The deployment of multicasting has been delayed partly because of the scalability issues of the related forwarding state. In unicast, address aggregation coupled with hierarchical address allocation has helped to achieve scalability. This can not be easily done for multicasting, since a multicast address corresponds to a logical group and does not convey any information on the location of its members. A multicast distribution tree requires all tree nodes to maintain per-group (or even per-group/source) forwarding state, and the number of forwarding state entries grows with the number of "passing-by" groups. As

---

[*] This material is based upon work supported by the National Science Foundation under Grant No. 9805436 and No. 9985195, CISCO/CORE fund No. 99-10060, DARPA award N660001-00-1-8936, TCS Inc., and DIMI matching fund DIM00-10071.

J. Crowcroft and M. Hofmann (Eds.): NGC 2001, LNCS 2233, pp. 172–188, 2001.

multicast gains widespread use and the number of concurrently active groups grows, more and more forwarding state entries will be needed. More forwarding entries translate into more memory requirements, and may also lead to slower forwarding process since every packet forwarding involves an address look-up. This perhaps is the main scalability problem with IP multicast when the number of simultaneous on-going multicast sessions is very large.

Recognition of the forwarding-state scalability problem has prompted some recent research in forwarding state reduction. Some architectures aim to completely eliminate multicast state at routers [5,9] using network-transparent multicast, which pushes the complexity to the end-points. Some other schemes attempt to reduce forwarding state by tunneling [11] or by forwarding-state aggregation [8,10]. Apparently, less entries are needed at a router if multiple forwarding state entries can be aggregated into one. Thaler and Handley analyze the aggregatability of forwarding state in [10] using an input/output filter model of multicast forwarding. Radoslavov et al. propose algorithms to aggregate forwarding state and study the bandwidth-memory tradeoff with simulations in [8]. Both these works attempt to aggregate routing state after the distribution trees have been established.

We propose a novel scheme to reduce multicast state, which we call *aggregated multicast*. The difference with previous approaches is that we force multiple multicast groups to share one distribution tree, which we call an *aggregated tree*. This way, the number of trees in the network may be significantly reduced. Consequently, forwarding state is also reduced: core routers only need to keep state per aggregated tree instead of per group. The trade-off is that this approach may waste extra bandwidth to deliver multicast data to non-group-member nodes. Simulation results demonstrate that, the more bandwidth we sacrifice, the more state reduction we can achieve. The management policy and functional requirements can determine the right point in this trade-off. In our earlier work [4], we introduced the basic concepts of aggregated multicast. In this paper, we propose an algorithm to assign multicast groups to delivery trees with controllable bandwidth overhead. We also propose a model to capture the membership patterns of multicast users, which can affect our ability to aggregate groups. Finally, we study the trade-off between aggregation versus bandwidth overhead using series of simulations.

The rest of this paper is organized as follows. Section 2 introduces the concept of aggregated multicast and discusses some related issues. Section 3 then formulates the tree sharing problem and presents an intuitive solution, and Section 4 provides a simulation study of our algorithm and cost/benefit evaluation. Finally Section 5 discusses the implications and contributions of our work.

2   Aggregated Multicast

Aggregated multicast is targeted as an intra-domain multicast provisioning mechanism. The key idea of aggregated multicast is that, instead of constructing a tree for each individual multicast session in the core network (backbone), one can force multiple multicast sessions share a single aggregated tree.

2.1   Concept

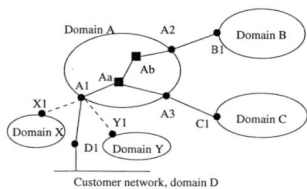

Customer network, domain D

Fig. 1. Domain peering and a cross-domain multicast tree, tree nodes: D1, A1, Aa, Ab, A2, B1, A3, C1, covering group G_0 (D1, B1, C1).

Fig. 1 illustrates a hierarchical inter-domain network peering. Domain A is a regional or national ISP's backbone network, and domain D, X, and Y are customer networks of domain A at a certain location (say, Los Angeles). Domain B and C can be other customer networks (say, in New York) or some other ISP's networks that peer with A. A multicast session originates at domain D and has members in domain B and C. Routers D1, A1, A2, A3, B1 and C1 form the multicast tree at the inter-domain level while A1, A2, A3, Aa and Ab form an intra-domain sub-tree within domain A (there may be other routers involved in domain B and C). The sub-tree can be a PIM-SM shared tree rooted at an RP (Rendezvous Point) router (say, Aa) or a bi-directional shared CBT (Center-Based Tree) tree centered at Aa or maybe an MOSPF tree. Here we will not go into intra-domain multicast routing protocol details, and just assume that the traffic injected into router A1 by router D1 will be distributed over that intra-domain tree and reaches router A2 and A3.

Consider a second multicast session that originates at domain D and also has members in domain B and C. For this session, a sub-tree with exactly the same set of nodes will be established to carry its traffic within domain A. Now if there is a third multicast session that originates at domain X and it also has members in domain B and C, then router X1 instead of D1 will be involved, but the sub-tree within domain A still involves the same set of nodes: A1, A2, A3, Aa, and Ab. To facilitate our discussions, we make the following definitions. We call **terminal nodes** the nodes where traffic enters or leaves a domain, A1, A2, and A3 in our example. We call **transit nodes** the tree nodes that are internal to the domain, such as Aa and Ab in our example. Using the terminology commonly used in DiffServ [2], terminal nodes are often *edge* routers and transit nodes are often *core* routers in a network.

In conventional IP multicast, all the nodes in the above example that are involved within domain A must maintain separate state for each of the three groups individually though their multicast trees are actually of the same "shape". Alternatively, in the aggregated multicast, we can setup a pre-defined tree (or establish on demand) that covers nodes A1, A2 and A3 using a single multicast group address (within domain A). This tree is called an **aggregated tree** (AT) and it is shared by more than one multicast groups. We say an aggregated tree T **covers** a group G if all terminal nodes for G are member nodes of T. Data from a specific group is encapsulated at the incoming terminal node. It is then distributed over the aggregated tree and decapsulated at exiting terminal nodes to be further distributed to neighboring networks. This way, transit router Aa and Ab only need to maintain a single forwarding entry for the aggregated tree regardless how many groups are sharing it. Furthermore, the use of aggregated multicast in one domain is transparent to the rest of the network.

2.2   Discussion

Aggregation reduces the required multicast state in a straightforward way. Transit nodes don't need to maintain state for individual groups; instead, they only maintain forwarding state for a smaller number of aggregated trees. On a backbone network, core nodes are the busiest and often they are transit nodes for many "passing-by" multicast sessions. Relieving these core nodes from per-micro-flow multicast forwarding enables better scalability with the number of concurrent multicast sessions.

The management overhead for the distribution trees is also reduced. First, there are fewer trees that exchange refresh messages. Second, tree maintenance can be a much less frequent process than in conventional multicast, since an aggregated tree has a longer life span. The control overhead reduction improves the scalability of multicast in an indirect yet important way.

The problem of matching groups to aggregated trees hides several subtleties. The set of the group members and the tree leaves are not always identical. A match is a **perfect** or **non-leaky match** for a group if all the tree leaves are terminal nodes for the group, thus traffic will not "leak" to any nodes that are not group members. For example, the aggregated tree with nodes (A1, A2, A3, Aa, Ab) in Fig. 1 is a perfect match for our early multicast group G_0 which has members (D1, B1, C1). A match may also be a **leaky match**. For example, if the above aggregated tree is also used for group G_1 which only involves member nodes (D1, B1), then it is a leaky match since traffic for G_1 will be delivered to node A3 (and will be discarded there since A3 does not have state for that group). A disadvantage of leaky match is that some bandwidth is wasted to deliver data to nodes that are not members for the group (e.g., deliver multicast packets to node A3 in this example). Leaky match may be unavoidable since usually it is not possible to establish aggregated trees for all possible group combinations.

Aggregated multicast can be deployed incrementally and it can interoperate with traditional multicast. First, we can invoke aggregation on a need-to basis. For example, we can aggregate only when the number of groups in a domain goes over a threshold. We can also choose to aggregate only when the aggregation causes reasonable bandwidth overhead, as we will discuss in detail in later sections. Second, aggregated multicast can co-exist with traditional multicast. Finally, the aggregation happens only within a domain, while it is transparent to the rest of the network including neighboring domains.

A related motivation for aggregated multicast is how to simplify the provisioning of multicast with QoS guarantees in future QoS-enabled networks. Regarding QoS support, per-flow-based traffic management requirement of Integrated Services [3] does not scale. Today people are backing away from it and are moving towards aggregated flow based Differentiated Services [2]. The intrinsic per-flow nature of multicast may be problematic for DiffServ networks especially in provisioning multicast with guaranteed service quality. Aggregated multicast can simplify and facilitate QoS management for multicast by pre-assignment of resource/bandwidth (or reservation on demand) in a smaller number of shared aggregated trees.

It is worth pointing out that our approach of "group aggregation" is fundamentally different from the "forwarding-state aggregation" approaches in [8,10]. We force multiple multicast groups to share a single tree, while their approach is to aggregate multiple multicast forwarding entries on each router locally. In a nutshell, we first aggregate and then route, while they first route and then aggregate. Note that the two approaches can co-exist: it is possible to further reduce multicast state using their approaches even after deploying our approach.

3   The Tree Sharing Problem

To implement aggregated multicast, two main problems must be worked out first: (1) what are the aggregated trees that should be established; (2) which aggregated tree is to be used for a certain group. In this section, we will formulate the tree sharing problem and propose a simple and intuitive algorithm; in the next section, we will present simulation results.

3.1   Why the Problem?

If aggregated multicast is only used by an ISP to provide multi-point connections among several routers that have heavy multicast traffic or that are strategically placed to carry inter-domain multicast, then a few number of trees can be pre-established and the matching (from group to a tree) is straightforward. The situation becomes complicated if aggregated multicast is used to a greater extend and the network is large.

Given a network with n edge nodes (nodes that can be terminal nodes for a multicast group), the number of different group combinations is about 2^n ($=C_n^2 + C_n^3 + ... + C_n^n = 2^n - 1 - n$, given that a group has at least two members), which grows exponentially with n. For a reasonably large network, it doesn't make sense to establish pre-defined trees for all possible groups – that number can be larger than the number of possible concurrently active groups. So we should and can only establish a subset of trees out of all possible combinations. This is where **leaky match** comes into play. Meanwhile, if aggregated multicast is used as a general multicast provisioning mechanism, then it becomes a necessity to dynamically manage and maintain aggregated trees since a static set of trees may not be very resource efficient all the time as groups come and go. For any solution one may have, the question is how much aggregation it can achieve and how efficient it is regarding bandwidth use.

3.2   Aggregation Overhead

A network is modeled as an undirected graph $G(V, E)$. Each edge (i, j) is assigned a positive cost $c_{ij} = c_{ji}$ which represents the cost to transport unit traffic from node i to node j (or from j to i). Given a multicast tree T, total cost to distribute a unit amount of data over that tree is

$$C(T) = \sum c_{ij}, \; link \; (i, j) \in T. \tag{1}$$

If every link is assumed to have equal cost, tree cost is simply $C(T) = |T| - 1$, where $|T|$ denotes the number of nodes in T.

Given a multicast group g and a tree T, we say tree T **covers** group g if all members of g are *in-tree nodes* of T (i.e., in the vertex set of T). If a group g is covered by a tree T, then any data packets delivered over T will reach all members of g, assuming a bi-directional tree. In a transport network, members of g are not necessarily **group member routers** (i.e., designated router with hosts in its subnet as group members), but rather they are edge routers connecting to other in-tree routers in neighboring domains.

Now consider a network in which routing algorithm A is used to setup multicast trees. Given a multicast group g, let $T_A(g)$ be the multicast tree computed by the routing algorithm. Alternatively, this group can be covered by a aggregated tree $T(g)$, **aggregation overhead** is defined as

$$\Delta(T, g) = C(T(g)) - C(T_A(g)). \tag{2}$$

Aggregation overhead directly reflects bandwidth waste if tree $T(g)$ is used to carry data for group g instead of the conventional tree $T_A(g)$ with encapsulation overhead not counted; i.e., bandwidth waste can be quantified as $D_g \times \Delta(T, g)$ if the amount of data transmitted is D_g. Note that, $T_A(g)$ is not necessarily the minimum cost tree (Steiner tree). Therefore, the aggregated tree $T(g)$ may happen to be more efficient than $T_A(g)$, and thus it is possible for $\Delta(T, g)$ to be negative.

3.3   Two Versions of the Problem

Static Pre-Defined Trees. In this version of the problem, we are given: a network $G(V, E)$, tree cost model $C(T)$, a set of N multicast groups, and a number n ($N >> n$). The goal is to find n trees (each of them covers a different node set) and a matching from a group g to a tree $T(g)$ such that every group g is covered by a tree $T(g)$, with the objective of minimizing total aggregation overhead. This is the problem we need to solve to build a set of pre-defined aggregated trees based on long-term traffic measurement information.

In reality, different groups may require different bandwidth and have different life time. Eventually they transmit different amounts of data (to all members, assumed). Aggregation overhead would be $D_g \times \Delta(T, g)$ for group g which transmits D_g amount of data. However, if D_g is independent of group size and group membership, then statistically the end effect will be the same if all groups are treated as if they have the same amount of data to deliver. Then the total aggregation overhead is simply $\sum_g \Delta(T, g)$. An **average percentage overhead** can be defined as

$$\delta_A = \frac{\sum_g \Delta(T, g)}{\sum_g C(T_A(g))} = \frac{\sum_g C(T(g))}{\sum_g C(T_A(g))} - 1. \tag{3}$$

Dynamic Trees. The dynamic version of the problem is more meaningful for practical purposes. In this case, instead of a static set of groups, groups dynamically come and go. Our goal is to find a procedure to generate and maintain (establish, modify and tear down) a set of trees and map a group to a tree when the group starts, while minimizing the percentage aggregation overhead.

If an upper bound is put on the number of trees that are allowed simultaneously, apparently the procedure in the dynamic tree matching problem can be used to solve the static tree matching problem: the given set of (static) groups are brought up one by one (without going down) and the dynamic tree matching procedure is used to generate trees and do the mapping; the resulting set of trees and the corresponding mapping are the solution (for the static tree sharing problem).

In the static case, the number of groups given is finite and assumed to be N. In the dynamic case, similarly we can specify N to be the average or maximum number of concurrently active groups. In both the static and dynamic problems, N and n (number of trees allowed) affect aggregation overhead. Intuitively, the closer n to N, the smaller the overhead will be. When $n = N$, the overhead can be 0 since each group can be matched to the tree computed by the routing algorithm. The question is if we can achieve meaningful aggregation ($N >> n$) while bandwidth overhead is reasonable.

3.4 Dynamic Tree Sharing with Aggregation Overhead Threshold Control ·

Here we present a solution for the dynamic tree sharing problem with the percentage aggregation overhead statistically controlled under a given threshold.

First we introduce some notations and definitions. Let MTS (multicast tree set) denote the current set of multicast trees established in the network. Let $G(T)$ be the current set of active groups covered by tree $T \in MTS$. Both MTS and $G(T)$ evolve with time, and the time parameter is implied but not explicitly indicated. For each $T \in MTS$, an average aggregation overhead for T is kept as

$$\tilde{\delta}(T) = \frac{\sum_{g \ in \ G(T)} \Delta(T,g)}{\sum_{g \ in \ G(T)} C(T_A(g))} = \frac{|G(T)| \times C(T)}{\sum_{g \ in \ G(T)} C(T_A(g))} - 1, \qquad (4)$$

and is updated every time $G(T)$ changes. $|G(T)|$ denotes the rank of set $G(T)$(or, the number of groups covered by T). At a certain time t, the average aggregation overhead for all groups $\delta_A(t)$(δ_A at time t) can be computed from $\tilde{\delta}(T,t)$($\tilde{\delta}(T)$ at time t). If tree T is used to cover group g, the percentage aggregation overhead for this match is

$$\delta(T,g) = \frac{C(T) - C(T_A(g))}{C(T_A(g))}. \qquad (5)$$

Let $\tilde{\delta}'(T,g)$ denote the average aggregation overhead if group g is covered by T and is to be added to $G(T)$, then

$$\tilde{\delta}'(T,g) = \frac{(|G(T)|+1) \times C(T)}{\sum_{g_x \ in \ \{G(T),g\}} C(T_A(g_x))} - 1 = \frac{(|G(T)|+1) \times C(T)}{\frac{|G(T)| \times C(T)}{1+\tilde{\delta}(T)} + C(T_A(g))} - 1. \qquad (6)$$

When a new multicast session goes on with group member set g, there are three options to accommodate this new group: (1) an existing tree covers g and is used to distribute packets for this new session; (2) an existing tree is extended to cover this group; (3) a new tree is established for this group.

Let b_t be the given bandwidth overhead threshold. The goal is to control (statistically) the total percentage overhead to be $\tilde{\delta}(T) < b_t$, for each $T \in MTS$, or $\delta_A < b_t$ (which is a weaker requirement).The following procedure determines how one of the above options is used:

(1) compute the "native" multicast tree $T_A(g)$ for g(e.g., using shortest-path tree algorithm as in MOSPF);

(2) for each tree T in MTS, if T covers g, compute $\tilde{\delta}'(T,g)$; otherwise compute an extended tree T^e to cover g and then compute $\tilde{\delta}'(T^e,g)$; if $\tilde{\delta}'(T,g) < b_t$ or $\tilde{\delta}'(T^e,g) < b_t$, then T or T^e is considered to be a candidate (to cover g);

(3) among all candidates, choose the one such that $C(T)$ or $C(T^e)+|G(T)| \times (C(T^e)-C(T))$ is minimum, denote is as T_m; T_m is used to cover g, update MTS (if T_m is an extended tree), $G(T_m)$, and $\tilde{\delta}(T_m)$;

(4) if no candidate found in step (2), $T_A(g)$ is used to cover g and is added to MTS and correspondingly $G(T_A(g))$ and $\tilde{\delta}(T_A(g))$ are recorded.

To extend tree T to cover group g (step (2)), a greedy strategy similar to Prim's minimum spanning algorithm [6] can be employed to connect T to nodes in g that are not covered, one by one.

Since each group has a limited life time, it will not be using a tree after that. A simple clean-up procedure can be applied when a group goes off: when a multicast session g goes off, g is removed from $G(T)$ where T is the tree used to cover g; if $G(T)$ becomes empty, then T is removed from MTS; T is pruned recursively for nodes no longer needed in the tree; $G(T)$ and $\tilde{\delta}(T)$ are updated. A node is no longer needed in tree T if it is a leaf and is not a member of any group $g \in G(T)$.

In the above algorithm description, we have assumed tree T is a bi-directional tree so that it can be used to cover any group whose members are all in-tree nodes of T. Apparently we can enforce that each tree is source-specific and each group needs to specify a source node, and the above algorithm still applies except that we may turn out to have more trees.

Bandwidth-Aware Aggregation. In all the aggregation overhead definitions we had above, bandwidth requirement of a multicast session is not considered. This is in agreement with today's IP multicast routing architecture where a group's bandwidth requirement is unknown to the routing protocols. At the same time, we assumed both bandwidth requirement and lifetime of a group are independent of the group size and member distribution. If bandwidth requirement is given for each multicast session (e.g., in future networks with QoS guarantee), the above algorithms can be extended to consider the bandwidth in a straightforward way. Due to space limitation, we do not present this formulation here.

3.5   Performance Metrics

We use the following metrics to quantify the effectiveness of an aggregation method.

Let $N(t)$ be the number of active multicast groups in the network at time t and $M(t)$ the number of trees, **aggregation degree** is defined as

$$AD(t) = \frac{N(t)}{M(t)}. \tag{7}$$

AD is an important indication of tree management overhead reduction. For example, the number of trees that need periodical refresh messages to keep state is reduced from N to $\frac{N}{AD}$.

Average Aggregation Overhead is

$$\delta_A(t) = \frac{\sum_g C(T(g))}{\sum_g C(T_A(g))} - 1 = \frac{\sum_{T \in MTS} |G(T)| \times C(T)}{\sum_{T \in MTS} \frac{|G(T)| \times C(T)}{1 + \bar{\delta}(T)}} - 1, \tag{8}$$

as defined in last subsection. δ_A reflects the extra bandwidth wasted to carry multicast traffic using shared aggregated trees.

Without loss of generality, we assume that a router needs one routing entry per multicast address in its forwarding table. Here we care about the **total number** of state entries that are installed at **all** routers involved to support a multicast group in a network. In conventional multicast, the total number of entries for a group equals the number of nodes $|T|$ in its multicast tree T (or subtree within a domain, to be more specific) – i.e., each tree node needs one entry for this group. In aggregated multicast, there are two types of state entries: entries for the shared aggregated trees and group-specific entries at terminal nodes. The number of entries installed for an aggregated tree T equals the number of tree nodes $|T|$ and these state entries are considered to be **shared** by **all groups** using T. The number of group-specific entries for a group equals the number of its terminal nodes because only these nodes need group-specific state.

Furthermore, we also introduce the concept of **irreducible state** and **reducible state**: group-specific state at terminal nodes is **irreducible**. All terminal nodes need such state information to determine how to forward multicast packets received, no matter in conventional multicast or in aggregated multicast. For example, in our early example illustrated by Fig. 1, node A1 always needs to maintain state for group G_0 so it knows it should forward packets for that group received from D1 to the interface connecting to Aa and forward packets for that group received from Aa to the interface connecting to node D1 (and not X1 or Y1), assuming a bi-directional inter-domain tree.

Given a set of groups $\mathcal{G}$, if each group g is serviced by a tree $T_A(g)$, then the total number of state entries is

$$N_A = \sum_{g \in \mathcal{G}} |T_A(g)|. \tag{9}$$

Alternatively, if the same set of groups are serviced using a set of aggregated trees MTS, the total number of state entries is

$$N_T = \sum_{T \in MTS} |T| + \sum_{g \in \mathcal{G}} |g|, \tag{10}$$

where $|T|$ is the number of nodes in T, and $|g|$ is the group size of g. The first part of (10) represents the number of entries to maintain for the aggregated trees, while the second part denotes the number of entries that source and exit nodes of a group need to maintain in order to determine how to forward and handle multicast data packets. **Overall state reduction ratio** can be defined as

$$r_{as} = 1 - \frac{N_T}{N_A} = 1 - \frac{\sum_{T \in MTS} |T| + \sum_{g \in \mathcal{G}} |g|}{\sum_{g \in \mathcal{G}} |T_A(g)|}. \tag{11}$$

A better reflection of state reduction achieved by our "group aggregation" approach, however, is the **reducible state reduction ratio**, which is defined as

$$r_{rs} = 1 - \frac{\sum_{T \in MTS} |T|}{\sum_{g \in \mathcal{G}} (|T_A(g)| - |g|)}; \tag{12}$$

i.e., the total number of entries needed to be maintained by transit nodes has been reduced from $\sum_{g \in \mathcal{G}} (|T_A(g)| - |g|)$ to $\sum_{T \in MTS} |T|$.

Another metric called **hit ratio** is defined as

$$HR(t) = \frac{number\ of\ groups\ covered\ by\ existing\ trees}{total\ number\ of\ groups} = \frac{N_h(t)}{N_t(t)}. \tag{13}$$

Both N_h and N_t are accumulated from time 0 (start of simulation) to time t. The higher $HR(t)$, the less often new trees have to be setup to accommodate new groups. Similarly **extend ratio** is defined as

$$ER(t) = \frac{number\ of\ groups\ covered\ by\ extended\ trees}{total\ number\ of\ groups} = \frac{N_e(t)}{N_t(t)}. \tag{14}$$

The "cost" to extend an existing is expected to be lower than setting-up a new tree. Percentage of groups that require to establish new trees is $1 - HR - ER$, up to time t.

4   Simulation Studies

In this section, we evaluate our approach by studying the trade-off between aggregation and bandwidth overhead using simulations. We find that we can achieve significant aggregation for reasonable bandwidth overhead. We attempt to test our approach in a wide variety of scenarios. Given the absence of large scale real multicast traces, we are forced to develop membership models that exhibits a locality and correlated group preferences.

4.1   Multicast Group Models

The performance of any aggregation is substantially affected by the distribution of multicast group members in the network. Currently multicast is not widely deployed and its usage has been limited, thus, trace data from real multicast sessions is limited and can only be considered as an indication of multicast patterns in large scale multicast. We develop and use the following different models to generate multicast groups.

In most multicast routing research literature, members of a multicast group are randomly chosen among all nodes. In this model, not only group members are assumed to be uncorrelated, but all nodes are treated the same as well. This well reflects member distribution of many applications, such as Internet gaming, but not all of them. In some applications, members tend to cluster together; for example, an Internet broadcast of Laker's basket ball game is watched by its local fans around Los Angeles area. Inter-group correlation is also an important factor; for example, members of a multicast group might tend to be in another group as well [12]. Neither does this model reflect the fact that not all nodes in the network are equivalent. For example, consider two nodes in MCI's backbone network: one is in Los Angeles and the other one is in Santa Barbara. It is very likely that the LA node has much more multicast sessions going through it than that of the Santa Barbara node given that MCI has a much larger customer base in LA. Besides, there can be historic correlation of group distribution among network nodes as well. For example, a customer company of MCI's Internet service has three locations in LA, Seattle and Houston, which are connected through MCI's backbone. There are often video conferences among these three sites, and when there is one, MCI's routers at the three places will be in a multicast session.

From the above discussions, to model multicast group distribution, the following factors have to be considered: (1) member distribution within a group (e.g., spread or clustered); (2) inter-group correlation; (3) node difference in multicast participation; (4) inter-node correlation; (5) group size distribution; i.e, how often we tend to have very small groups or very large groups. Factor (5) has been not discussed above, but clearly it is very important as well. Several models are described in [12] where factors (1) and (2) are considered. The terms of affinity and disaffinity are used in [7] to describe the clustering and spreading out tendencies of members within a group.

In our work, we use the **node weighted** framework which incorporates the difference among network nodes (factor (3)). In this framework, each node is assigned a weight representing the probability for that node to be in a group. We have two models to generate groups based on node weight assignment, which gives rise to two different models.

The Random Node-Weighted Model. This model statistically controls the number of groups a node will participate based on its weight: for two nodes i and j with weight $w(i)$ and $w(j)$ $(0 < w(i), w(j) \leq 1)$, let $N(i)$ =the number of groups that have i as a member and $N(j)$ =the number of groups that have j as a member, then $\frac{N(i)}{N(j)} = \frac{w(i)}{w(j)}$ in average. Assuming the number of nodes in the network is N and nodes are numbered from 1 to N. For each node i, $1 \leq i \leq N$, it is assigned a weight $w(i)$, $0 \leq w(i) \leq 1$. Then a group can be generated as the following procedure:

> **for** $i = 1$ *to* N **do**
>> generate a random number between 0 and 1, let it be p
>> **if** $p < w(i)$ **then**
>>> add i as a group member
>> **end if**
> **end for**

The Group-Size Controlled Model. In this model, we want to have more accurate control over the size of the groups we generate. For this reason, we use the following procedure to generate groups with a given group size that follows a given probability mass function pmf $p_X(x)$:

> generate group size n according to $p_X(x)$
> **while** the number of member is less than n **do**
>> randomly pick up a non-member, let it be i
>> generate a random number between 0 and 1, let it be p
>> **if** $p < w(i)$ **then**
>>> add i as a group member
>> **end if**
> **end while**

This model controls the group-size distribution; however, nodes no longer participate in groups according to their weights (i.e., we no longer have $\frac{N(i)}{N(j)} = \frac{w(i)}{w(j)}$ in average).

4.2 Simulation Results

We present results from simulation using a network topology abstracted from a real network topology, AT&T IP backbone [1], which has a total of 123 nodes: 9 gateway routers, 9 backbone routers, 9 remote GSR (gigabit switch router) access router, and 96 remote access routers.

The abstract topology is constructed as follows. First, we "contract" all the attached remote access routers of a gateway router or a backbone router into one node (connecting to the original gateway/backbone router), which is called a **contracted node**. Since a gateway router in the backbone represents connectivity to other peering network(s) and/or Internet public exchange point(s), a neighbor node called **exchange node** is added to each gateway router to represent such external connectivity. The result is a simplified network with 54 nodes. Among these nodes, gateway nodes (9 of them) and backbone nodes (9 of them) are assumed to be *core routers* only (i.e., will not be terminal nodes for any multicast group) and are assigned weight 0. Each access router is assigned weight 0.01, and a "contracted" node's weight is the summation of the weights of all access routers from which it is contracted. Exchange nodes are assigned weight ranging from 0.1 to 0.9 in different simulation runs.

In simulation experiments, multicast connection requests arrive as a Poisson process with arrival rate λ. Each time a connection comes up, all group members are specified. Group membership is generated using the node weighted framework discussed in Section 4.1. Connections' life time has a Poisson distribution with average μ. At steady state, average number of connections is $\bar{N} = \lambda \times \mu$. The algorithm specified in last section is used to establish/mantain trees and map a group to a tree. The routing algorithm A is shortest-path tree algorithm with root randomly chosen from all group members. Performance data is collected at certain time points (e.g., at $T = 10\mu$), when stead state is reached, as "snapshot".

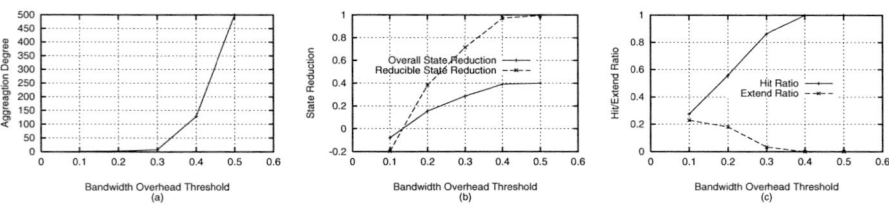

Fig. 2. Aggregation vs. bandwidth overhead threshold, groups are generated using group-size-controlled model.

In our first experiment, an exchange node is assigned a weight 0.25 or 0.9 according to link bandwidths of the original gateway– the rationale is that, the more the bandwidth on the outgoing (and incoming) links of a node, the more the number of multicast groups it may participate. Groups are generated using the group-size-controlled model. Group size is uniformly distributed from 2 to 36 and the average number of concurrently active groups is $\lambda\mu = 1000$. Fig. 2 shows the results of aggregation degree (a), state reduction (b) and hit/extend ratios (c) vs. bandwidth overhead threshold. We can see that, aggregation degree increases as the bandwidth threshold is increased – if we are willing sacrifice more bandwidth, we can accommodate more multicast groups into a shared aggregated tree. Apparently this agrees with our intuition. Fig. 2(b) shows that, overall state reduction ratio and reducible state ratio also increase with bandwidth overhead threshold – as we squeeze more groups into an aggregated tree, we need fewer trees and achieve more state reduction. Fig. 2(c) tells us that, hit ratio goes up and extend ratio goes down with increasing threshold. This is consistent with the trend of other metrics (aggregation degree and state reductions). When more groups can share an aggregated tree, it is more likely for an incoming group to be covered by an existing tree and thus it becomes less often to setup new trees or extend existing trees.

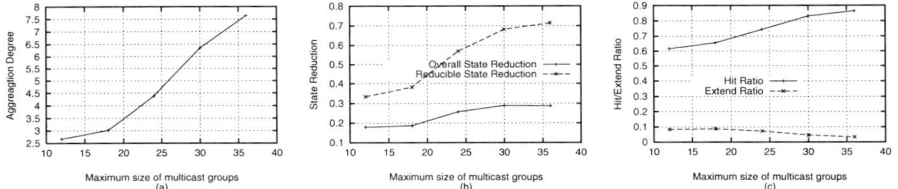

Fig. 3. Aggregation vs. maximum size of multicast groups, groups are generated using group-size-controlled model.

In our second experiment, we keep the bandwidth overhead threshold at a fixed value (=0.3 for results presented here) and vary the upper bound of group size (still uniformly distributed with lower bound 2) while keep all other parameters the same as in the first experiment. The results in Fig. 3 demonstrate the effect of group size on aggregation: if there are more larger groups, then we can aggregate more groups into sharing trees. As groups become larger, so do their multicast trees. A larger tree can "cover" more groups than a smaller one under the same overhead threshold (i.e., there are more subtrees of a larger tree within that threshold).

We want to see how node weight affects aggregation. Here we also keep the bandwidth overhead threshold at a fixed value (=0.3 for results presented here) and group size is uniformly distributed from 2 to 36. All other parameters are the same as in our first experiment, while we vary the weight of *exchange nodes*

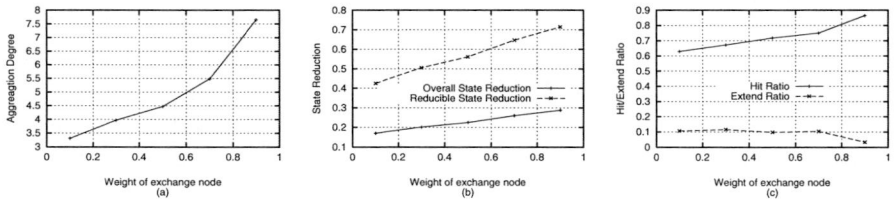

Fig. 4. Aggregation vs. weight of exchange nodes, groups are generated using group-size-controlled model.

from 0.1 to 0.9. The results are shown in Fig. 4. The higher the weight of a node, the larger the number of groups it may participate. As we increase the weights of those *exchange nodes*, multicast groups are more likely to "concentrate" on these nodes, and better aggregation is achieved as the results show.

The same set of experiments are also conducted using the random node-weighted model. In Fig. 5, we plot the results of an experiment similar to our first one. The results demonstrate similar trends, although the actual values differ. Note that the state reduction seems to be comparable.

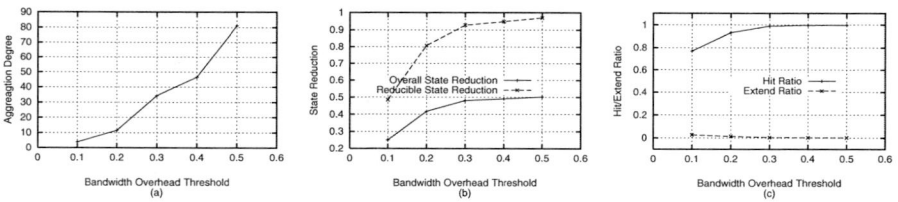

Fig. 5. Aggregation vs. bandwidth overhead threshold, groups are generated using random model.

We also examine how the aggregation scales with the (statistic average) number of concurrent groups. We run simulations with different $\lambda\mu$ products while keep all other parameters fixed. Fig. 6 plots the results for two different bandwidth overhead thresholds. It is no surprise that as more groups are pumped into the network, the aggregation degree increases – in average, more groups can share an aggregated tree. The scaling trend is encouraging: as the average number of (concurrent) groups is increased from 1000 to 9000, the number of aggregated trees is increased from 29 to 46 only (with bandwidth overhead threshold 0.3). At the same time, reducible state reduction ratio is getting close to 1.

To summarize, we observe that the state reduction can be significant up to 50% for the overall state. Furthermore, we can get significant reduction even for

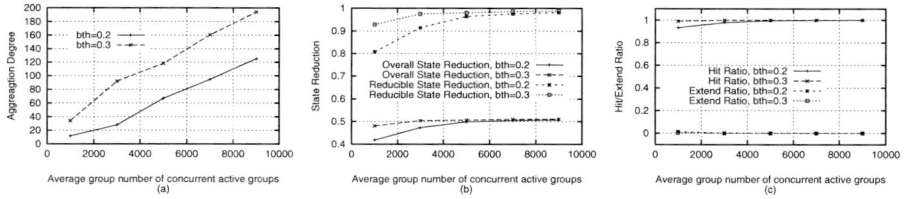

Fig. 6. Aggregation vs. concurrent group number, groups are generated using random model.

small bandwidth overhead. Finally, our approach has the right trend: the state-reduction increases as the number and the size of groups increase. This way, the aggregation becomes more effective when it is really needed.

We also have to warn the limitations of such simulation studies. As we have found, multicast group models (size distribution and member distribution as controlled by node weights) can significantly affect aggregation; thus how aggregated multicast is going to work out in real networks depends a lot on such factors in practice. Therefore, it is important to develop realistic multicast scenarios to evaluate any aggregation approach.

5   Conclusions and Future Work

We propose a novel approach to address the problem of multicast state scalability. The key idea of aggregated multicast is to force groups into sharing a single delivery tree. This comes in contrast to other forwarding-state aggregation approaches that first create multiple trees and then try to aggregate the state locally on each router. A key concept in our approach is that it sacrifices bandwidth to reduce the routing state.

Our work could be summarized in the following points:

- We introduce the concept of aggregated multicast and discuss several related issues.
- We formulate the tree sharing problem and present a simple and effective algorithm to establish aggregated trees and dynamically match groups with existing trees.
- We propose performance metrics that can be used to evaluate our approach.
- We show that our approach seems to be very promising in a series of simulation experiments. We can achieve significant state aggregation (up to 50%) with relatively small bandwidth overhead(10% to 30%).

Our work suggests that the benefits of aggregated multicast lie in the following two areas: (1) control overhead reduction by reducing the number of trees needed to be maintained in the network; (2) state reduction at core nodes. While the price to pay for that is bandwidth waste. Our simulation results confirm our claim while demonstrate the following trends: (1) as we are willing to sacrifice more bandwidth (by increasing the bandwidth overhead threshold), more or better aggregation is achieved; (2) better aggregation is achievable as the number and size of concurrent groups increases. The last is specially important since one basic goal of aggregated multicast is to achieve better scalability regarding the number of concurrent groups.

Future Work. Our scheme simplifies multicast management and could lend itself to a mechanism of QoS provisioning and multicast traffic engineering with the appropriate use of DiffServ or MPLS. We find that this by itself could be a sufficient motivation for studying aggregated multicast.

References

1. AT&T IP Backbone. http://www.ipservices.att.com/backbone/, 2001.
2. S. Blake et al. An architecture for differentiated services. *IETF RFC 2475*, 1998.
3. R. Braden, D. Clark, and S. Shenker. Integrated services in the internet architecture: An overview. *IETF RFC 1633*, 1994.
4. Aiguo Fei, Jun-Hong Cui, Mario Gerla, and Michalis Faloutsos. Aggregated multicast: an approach to reduce multicast state. *To appear in Sixth Global Internet Symposium(GI2001)*, November 2001.
5. P. Francis. Yoid: Extending the internet multicast architecture.
   http://www.aciri.org/yoid/docs/index.html.
6. U. Manber. *Introduction to Algorithms: A Creative Approach*. Addison-Wesley Publishing Company, 1989.
7. G. Philips and S. Shenker. Scaling of multicast trees: Comments on the chuang-sirbu scaling law. In *Proceedings of ACM SIGCOMM'99*, pages 41–51, September 1999.
8. P.I. Radoslavov, D. Estrin, and R. Govindan. Exploiting the bandwidth-memory tradeoff in multicast state aggregation. Technical report, USC Dept. of CS Technical Report 99-697 (Second Revision), July 1999.
9. Y. Chu S. Rao and H. Zhang. A case for end system multicast. In *Proceedings of ACM Sigmetrics 2000*, June 2000.
10. D. Thaler and M. Handley. On the aggregatability of multicast forwarding state. In *Proceedings of IEEE INFOCOM 2000*, Tel Aviv, Israel, March 2000.
11. J. Tian and G. Neufeld. Forwarding state reduction for sparse mode multicast communications. In *Proceedings of IEEE INFOCOM'98*, San Francisco, California, March 1998.
12. T. Wong and R. Katz. An analysis of multicast forwarding state scalability. In *Proceedings of the 8th International Conference on Network Protocols (ICNP)*, Japan, November 2000.

Tree Layout for Internal Network Characterizations in Multicast Networks*

Micah Adler, Tian Bu, Ramesh K. Sitaraman, and Don Towsley

Department of Computer Science
University of Massachusetts, Amherst, MA 01003
{micah,tbu,ramesh,towsley}@cs.umass.edu

Abstract. There has been considerable activity recently to develop monitoring and debugging tools for a multicast session (tree). With these tools in mind, we focus on the problem of how to lay out multicast sessions so as to cover a set of links of interest within a network. We define two variations of this layout (cover) problems that differ in what it means for a link to be covered. We then focus on the minimum cost problem, to determine the minimum cost set of trees that cover the links in question. We show that, with few exceptions, the minimum cost problems are NP-hard and that even finding an approximation within a certain factor is NP-hard. One exception is when the underlying network topology is a tree. For this case, we demonstrate an efficient algorithm that finds the optimal solution. We also present several computationally efficient heuristics and their evaluation through simulation. We find that two heuristics, a greedy heuristic that combines sets of trees with three or fewer receivers, and a heuristic based on our tree algorithm, both perform reasonably well. The remainder of the paper applies our techniques to the vBNS network and randomly generated networks, examining the effectiveness of the different heuristics.

1 Introduction

Multicast is a technology that shows great promise for providing the efficient delivery of content from a single source to many receivers. An interoperable networking infrastructure is nearly in place (PIM-SM/MSDP/MBGP,SSM) and the development of mechanisms for congestion control and reliable data delivery are well under way [4,9]. However, deployment of multicast applications lags behind, in large part because of a lack of debugging and monitoring tools. Recently, several promising approaches and protocols have been proposed for the purpose of aiding the network manager or the multicast application designer in this task. These include the use of end-to-end measurements for inferring internal behavior on a multicast tree [2], the development of the multicast route monitor (MRM) protocol [3], and a number of promising fault monitoring tools, [13,16]. All of these address the problem of identifying performance and/or fault behavior on a *single multicast tree*.

---

* This work is sponsored in part by the DARPA and Air Force Research Laboratory under agreement F30602-98-2-0238.

J. Crowcroft and M. Hofmann (Eds.): NGC 2001, LNCS 2233, pp. 189–204, 2001.

Although considerable progress has been made in developing tools for a single tree, little attention has been paid on how to apply these tools to monitor an entire network, or even a subset of the network. We address this problem; namely, given a set of links whose behavior is of interest, how does one choose a set of minimum cost multicast trees within the network on which to apply these tools so as to determine the behavior of the links in question? The choice of trees, of course, is determined by the multicast routing algorithm. This raises a related question, namely, does the multicast routing algorithm even allow a set of trees that that will allow one to determine the behavior of the links of interest. We refer to this latter problem as the Multicast Tree Identifiability Problem (MTIP) and the first problem as the Minimum cost Multicast Tree Cover Problem (MMTCP).

We refer to the behavior measurement of a link as a *link measure*. Note that solutions to MTIP and MMTCP depend on details of the mechanism used to determine the link measures. Consequently, we introduce two versions of these problems, the weak and the strong cover problems. Briefly, the weak cover problem is based on the assumption that it is sufficient that each link of interest appear in at least one tree. The strong cover problem requires that each link occur between two branching points in at least one tree.

Briefly, the paper makes the following contributions.

- We establish that the cover problems are NP-hard and that in some cases, finding an approximation within a certain factor of optimal is also NP-hard. Thus, we also propose several heuristics and show through simulation that a greedy heuristic that iteratively combines trees containing a small number of receivers performs reasonably well.
- We provide polynomial time algorithms that find optimal solutions for a restricted class of network topologies, including trees. This algorithm can be used to provide a heuristic for sparse, tree like networks. This heuristic is also shown through simulation to perform well.
- We apply our techniques to the vBNS network and randomly generated networks, examining the effectiveness of the different heuristics.

The remainder of the paper proceeds as follows. Section 2 presents the model for MTIP and MMTCP, as well as the two types of covers we consider. Section 3 introduces several approximation algorithms and heuristics for MMTCP. In Section 4 we present efficient algorithms that find the optimal MMTCP solution for the special case where the underlying network topology is a tree. Section 5 presents the results of simulation experiments on the VBNS network and randomly generated networks. Last, Section 6 concludes the paper.

2   Model and Assumptions

We represent a network N by a directed graph $N = (V(N), E(N))$ where $V(N)$ and $E(N)$ denote the set of nodes and links within N respectively. When unambiguous, we will omit the argument N. Our interest is in multicast trees embedded within N. Let $S \subseteq V(N)$ be a set of potential multicast senders, and $R \subseteq V(N)$ a set of potential multicast receivers. Let $T = (V(T), E(T))$ denote a directed (multicast) tree with a source $s(T)$ and a set of leaves $r(T)$. We require that $s(T) \in S$ and $r(T) \subseteq R$. Let A be a mapping that takes a source $s \in S$ and receiver set $r \subseteq R$ and returns a tree $A(s, r)$. In the context of a network, A corresponds to the multicast routing algorithm. Examples include DVMRP [12], and PIM-DM and PIM-SM [8]. Let $\mathcal{T}(A, S, R) = \{A(s, r) : s \in S, r \subseteq R \setminus \{s\}\}$, i.e., $\mathcal{T}(A, S, R)$ is the set of all possible multicast trees that can be embedded in N using multicast routing algorithm A. We shall henceforth denote $\mathcal{T}(A, S, R)$ by $\mathcal{T}(S, R)$, omitting the dependence on A.

We associate the following cost with a multicast tree $T \in \mathcal{T}(S, R)$,

$$C(T) = C_T^0 + \sum_{l \in E(T)} C_l, \tag{1}$$

where the first term is a "per tree cost" and the second is a "per link cost". For example, IP multicast requires each multicast router to maintain per flow state. This is accounted for by the per tree cost. The per link cost is the cost for sending probe packets through a link. The two problems of interest to us are as follows:

Multicast Tree Identifiability Problem. Given a set of multicast trees $\Psi \subseteq \mathcal{T}(S, R)$, and a set of links $L \subseteq E$, is L identifiable by the set of trees Ψ?

Minimum Cost Multicast Tree Cover Problem. Given $S, R \subseteq V$, $L \subseteq E$ and L is identifiable by $\mathcal{T}(S, R)$, what is the minimum cost subset of $\mathcal{T}(S, R)$ sufficient to cover L? In other words, find $\Psi \subseteq \mathcal{T}(S, R)$ that covers L and minimizes

$$C(\Psi) = \sum_{T \in \Psi} C(T)$$

We distinguish two types of solutions to both of these problems. These solutions differ in what exactly is meant by a cover. We say that a node v is a branch point in tree T if v is either a root or a leaf, or v has more than one child. A path $l = (v_1, v_2, \ldots, v_n)$ is said to be a *logical link* within T if v_1 and v_n are branch points, $v_2, \ldots, v_{n-1}$ are not, and $(v_i, v_{i+1}) \in E(T)$, $i = 1, \ldots, n - 1$.

- **Strong cover:** Given a set of trees Ψ, Ψ is the strong cover of link $l = (u, v)$ if there exists a $T \in \Psi$ such that both u and v are branch points in T. Ψ is the strong cover of a link set L if $\forall l \in L$, Ψ is the strong cover of l.
- **Weak cover:** Given a set of trees Ψ, Ψ is the weak cover of link l if there exists a $T \in \Psi$ such that $l \in E(T)$. We say that Ψ is the weak cover of a link set L if $\forall l \in L$, Ψ is the weak cover of l.

We refer to the problems of finding these types of solutions as *S-MTIP/S-MMTCP* and *W-MTIP/W-MMTCP* respectively. Several cases are of interest to us. One is where $L = E$, i.e., where the objective is to cover the entire network. A second is where L consists of one link, $|L| = 1$. If, $\forall l$, we set $C(l) = 0$, the problem becomes that of covering the link set L with the set of trees with minimum total per tree cost.

The solutions to *S-MTIP* and *W-MTIP* are straightforward and are found in [1]. Henceforth, $Cov(\Psi, COVER)$ is a function, described in [1], that returns the maximum set of links identifiable by the set of trees Ψ where $COVER$ can be either 'strong' or 'weak'.

In this paper, we assume that both the network topology and multicast routing algorithm are given. Most debugging and monitoring is performed by network operators. They either know the exact topology or can easily discover it. They then can apply the network multicast routing algorithm to obtain the set of possible trees. End-users often have access to topology information and can apply tools such as `mtrace` (e.g., [5]) to identify the set of possible trees. Our model doesn't account for temporary routing changes. However, according to measurements in [11], these occur very infrequently as compared to the duration of a debugging session. Moreover, once a routing change is detected (e.g. by running mtrace periodically), one can always re-apply our techniques on the new topology and obtain a new layout of multicast trees.

3   Approximating the Minimum Cost Multicast Tree Cover Problem

We have defined two types of Multicast Tree Cover Problems: the S-MMTCP and the W-MMTCP. Unfortunately, as the following theorem shows, not only are these problems NP-Complete, we cannot even expect to find a good quality approximation to these problems. In particular, we can demonstrate the following (the proof is found in [1]):

Theorem 1. *For each of S-MMTCP and W-MMTCP, it is NP-Hard to find a solution that is within a factor of $(1 - \epsilon) \ln |L|$ of the optimal solution, for any $\epsilon > 0$. These problems are also still NP-Hard even with the restriction that $\forall l, C(l) = 0$.*

Since we cannot expect to solve these problems exactly, in the remainder of this section, we focus on approximation algorithms and heuristics for good solutions. In Section 3.1, we focus on approximation algorithms for the case where the goal is to minimize the total cost of setting up the multicast trees. We provide polynomial time algorithms that have a provable bound on how close to optimal the resulting solution is for the S-MMTCP and the W-MMTCP. In Section 3.2, we describe extensions to these algorithms for the problem of approximating the general MMTCP. The resulting algorithms only run in polynomial time when the number of possible receivers is $O(\log |E|)$, and thus, to approximate the general MMTCP more efficiently, we also propose a heuristic that always finds a solution to the general MMTCP in polynomial time. In Section 5, we experimentally verify the quality of the solutions found by this heuristic.

3.1   Minimizing the Total Per-Tree Cost

We first study how to approximate the MMTCP when the goal is only to minimize the total cost for setting up the multicast trees but not the cost for multicast traffic to travel links, i.e., $C_l = 0$ in (1). This problem is simpler, since without a cost for link traffic, if a sender is performing a multicast, there is no additional cost for sending to every receiver. Thus, we can assume that any active sender multicasts to every possible receiver. Note, however, that by Theorem 1, even this special case is NP-Hard to approximate within better than a $\ln |L|$ factor. We describe algorithms for this problem that achieve exactly this approximation ratio. These algorithms rely on the fact that when $C(l) = 0, \forall l$, then both the W-MMTCP and the S-MMTCP can be solved using algorithms for the weighted Set-Cover problem, which is defined as follows:

- The **weighted Set-Cover** problem: given a finite set $B = \{e_1, e_2, ..., e_m\}$ and a collection of subsets of B, $\mathcal{F} = \{B_1, B_2, ..., B_n\}$, where each $B_i \subseteq B$ is associated with a weight w_i, find the minimum total weight subset $\hat{\mathcal{F}} = \{\hat{B}_1, ..., \hat{B}_k\} \subseteq \mathcal{F}$ such that each $e_i \in B$ is contained in some $\hat{B}_j \in \hat{\mathcal{F}}$.

To use algorithms for the weighted Set-Cover problem to solve the S-MMTCP or W-MMTCP, we simply set $B = L$, $B_i = \{l \in L$ such that l is in the cover (strong or weak, respectively) produced by T_i, where T_i is the tree produced by sender i multicasting to every receiver$\}$. The weight of B_i is the per-tree cost of multicasting from sender i. Any solution to the resulting instance of the weighted Set-Cover problem produces a S-MMTCP (W-MMTCP, resp.) solution of the same cost. Using this idea, we introduce two algorithms for the MMTCP: a greedy algorithm modeled after a weighted Set-Cover algorithm analyzed by Chvatal [6], and an algorithm that uses 0-1 integer programming, constructed using a weighted Set-Cover algorithm analyzed by Srinivasan [15].

Greedy Algorithm: The intuition behind the greedy algorithm is simple. Assume first that the per-tree cost is the same for every multicast tree. In this case, the algorithm, at every step, chooses the multicast tree that covers the most remaining uncovered links. This is repeated until the entire set of links is covered. When different trees have a different per-tree cost, then instead of maximizing the number of new links covered, the algorithm maximizes the number of new links covered, divided by the cost of the tree. Intuitively, this maximizes the "profit per unit cost" of the new tree that is added.

The details of the algorithm are shown in Figure 1. This algorithm is easily seen to run in polynomial time for all three types of covers. For the W-MMTCP and S-MMTCP, Theorem 2 provides a bound on how good an approximation the algorithm produces.

Theorem 2. *For any instance I of S-MMTCP or W-MMTCP with $C(l) = 0, \forall l$, the greedy algorithm finds a solution of cost at most $\left(\ln d + \frac{5}{8} + \frac{1}{2d}\right) \cdot OPT$, where $d = \min\left(|L|, \max_{i=1}^n |L_i|\right)$, and OPT is the cost of the optimal solution to I.*

We see from Theorems 1 and 2 that the performance of the greedy algorithm is the best that we can hope to achieve. However, these theorems only apply to the worst case performance; for the average case, the performance may be much better, and the best algorithm may be something completely different. We investigate this issue further by introducing a second approximation algorithm, based on 0-1 integer programming. We shall see in Section 5 that the 0-1 integer

1. Compute Ψ using the multicast routing protocol A where Ψ the set of all multicast trees from a source in S to every receiver in R. $\forall T_i \in \Psi$, set its cost $c_i = C^0(T_i)$.
2. Set $J = \Psi, \hat{J} = \emptyset, \hat{L} = \emptyset$.
3. If $L = \hat{L}$, then stop and output $\hat{J}$.
4. $\forall T_i \in J \setminus \hat{J}$, set $L_i = Cov(\hat{J} \cup \{T_i\}, COVER)$. Find $T_i \in J \setminus \hat{J}$ that maximizes $|(L_i \cap L) \setminus \hat{L}|/c_i$.
5. $\hat{L} = \hat{L} \cup (L_i \cap L)$, $\hat{J} = \hat{J} \cup \{T_i\}$. Go to step 3.

Fig. 1. The Greedy Algorithm to Approximate MMTCP.

programming algorithm performs better than the greedy algorithm in some cases. The details of this algorithm are omitted from this version of the paper due to space limitations and can be found in [1]; here we only state the following theorem that demonstrates how good a solution is provided by this approach. The proof of the theorem is also presented in [1].

Theorem 3. *For any instance I of either the S-MMTCP or the W-MMTCP, the 0-1 linear programming algorithm finds a solution of cost at most $OPT(1 + O(\max\{\ln(m/OPT), \sqrt{\ln(m/OPT)}\}))$, where OPT is the optimal solution to I.*

We also point out that in the Set-Cover problem, if we let $d = \max |B_i|$, then even for $d = 3$, the set cover problem is still NP-hard. However, it can be solved in polynomial time provided that $d = 1$ or $d = 2$. Since we can transform the S-MMTCP and the W-MMTCP to Set-Cover problems, we know that the S-MMTCP and the W-MMTCP can be solved in polynomial time given that $\max_i |L \cap E(T_i)| \leq 2$ where T_i is the tree produced by sender i multicasting to every receiver.

3.2   Minimizing the Total Cost

We next look at the general MMTCP, where the goal is to minimize the total cost. In addition to the per-tree cost of the multicast trees used in the cover, this includes the cost of multicast traffic traveling on the links used by the trees. Both the greedy algorithm and 0-1 integer programming algorithm can be extended to approximate the general problem. For the greedy algorithm, we simply replace the first step with:

1. Compute Ψ using the multicast routing protocol A. where Ψ is the set of all multicast trees from a source in S to any subset of R. $\forall T_i \in \Psi$, compute its cost $c_i = C(T_i)$.

The bound from Theorem 2 on the quality of approximation achieved by the greedy algorithm also applies to this more general algorithm. However, the greedy algorithm has a running time that is polynomial in $|\Psi|$. For the algorithm of Section 3.1, $|\Psi| = |S|$, which results in a polynomial time algorithm, but for the more general algorithm considered here, $|\Psi| = |S| \cdot (2^{|R|} - 1)$. Thus, the more general approximation algorithm only has a polynomial running time when $|R| = O(\log n)$, where n is the size of the input to the MMTCP problem (i.e., the description of N, S, R and L). The analogous facts also apply to the 0-1 integer programming algorithms.

In order to cope with large values of $|R|$ in the general MMTCP, we also introduce the fast greedy heuristic, which always runs in polynomial time. Fast greedy is like the greedy algorithm, except that instead of considering all possible multicast trees (i.e., every tree from a sender to a subset of the receivers), it restricts itself to only those trees that contain 3 receivers (or, in the case of a weak cover, 1 receiver). There will be at most a polynomial number of such trees. Fast greedy then uses the greedy strategy to choose a subset of these trees covering all required links, and then merges the trees with the same sender. The details of this heuristic are described in Figure 2. We shall see in Section 5 that the performance of the fast greedy heuristic is often close to that of the greedy algorithm.

1. If COVER = 'strong', apply the multicast routing protocol A to compute Ψ, the set of all multicast trees that have one sender in S and three receivers in R.
If COVER = 'weak', apply the multicast routing protocol A to compute Ψ, the set of all multicast trees (paths) that have one sender in S and one receiver in R.
$\forall T_i \in \Psi$, compute its cost $c_i = C(T_i)$.
2. For all $T_i \in \Psi$, set $E_i = E(T_i)$.
3. Set $J = \Psi, \hat{J} = \emptyset, \hat{L} = \emptyset$.
4. If $L = \hat{L}$, then aggregate all the trees in $\hat{J}$ who share the same source node and output $\hat{J}$. Stop.
5. For all $T_i \in J \setminus \hat{J}$, set $\Psi_i = \hat{J} \cup \{T_i\}$ and aggregate all the trees in Ψ_i sharing a common source as one tree. Set $L_i = Cov(\Psi_i, COVER)$.
Find $T_i \in J \setminus \hat{J}$ that maximizes $|(L_i \cap L) \setminus \hat{L}|/c_i$.
6. $\hat{L} = \hat{L} \cup (L_i \cap L), \hat{J} = \hat{J} \cup \{T_i\}$. For each $T_j \in J \setminus \hat{J}$, if T_j shares the same source with T_i, $E_j = E_j \setminus E_i$, $c_j = \sum_{l \in E_j} C(l)$. Go to step 4.

Fig. 2. Fast Greedy Heuristic to Approximate MMTCP.

4   Finding the Optimal Solution in a Tree Topology

We saw in Theorem 1 that we cannot hope to find an efficient algorithm that solves any of version of the MMTCP in general. However, this does not rule out the possibility that it is possible to solve these problems efficiently on certain classes of network topologies. In this section, we study the MMTCP in the case that the underlying network topology N is a tree. This is motivated by the hierarchical structure of real networks, which can be thought of as having a tree topology with a small number of extra edges. We shall see in Section 5 that algorithms for the tree topology can be adapted to provide a very effective heuristic for such hierarchical topologies. We use this heuristic to provide good solutions to the W-MMTCP problem for the topologies of the vBNS network, as well as the Abilene network.

Our algorithm for the tree topology is guaranteed to find the optimal solution in polynomial time. We shall describe this algorithm for the (easier) case of the W-MMTCP. In order to describe this algorithm more concisely, we shall make some simplifying assumptions. In particular, we assume that N is a rooted binary tree, that the per tree cost of every multicast tree is zero, that $C((a, b)) = C((b, a))$ and that the cover requirement on a link can be satisfied from either direction. For the W-MMTCP, these assumptions can be removed by making the algorithm slightly more complicated, but without significantly increasing the running time of the algorithm. For the S-MMTCP, all of the assumptions can be removed except for the assumption that N is a binary tree.

The algorithm uses the technique of dynamic programming, and starts by creating a table, with one row in the table for each link l of the tree, and $|S|^2$ entries in each row, labeled $C_{<l,i,j>}$, for $0 \leq i, j \leq |S|$. For link l connecting nodes $u, v \in L$, where u is closer to the root, let ST_l be the subtree rooted at node v, together with link l and node u. The value computed for entry $C_{<l,i,j>}$ is the minimum possible total cost for the tree ST_l (removed from the rest of the network), subject to the following conditions: all of the links that are required to be covered in N are covered in ST_l, u is a source that generates j multicast sessions that are routed across l, and u is also a receiver that receives i multicast sessions. If there are less than i senders in $ST_l - u$, or $j > 0$ and there are no receivers in $ST_l - u$, then we call the pair (i, j) *invalid* for link l, and the value of entry $C_{<l,i,j>}$ is set to infinity.

We compute the values in the table one row at a time, in decreasing order of the distance of the corresponding links from the root. When l is connected to a leaf of the tree N, it is straightforward to compute $C_{<l,i,j>}$ for all i, j, since if (i, j) is valid, then $C_{<l,i,j>} = (i + j)C_l$. We now show how to compute the remaining entries $C_{<l,i,j>}$ for a link l connecting nodes u and v, where u is closer to the root, and v is connected to two links m and n, as depicted in Figure 3. Since m and n are further from the root than l, we can assume that $C_{<m,i_m,j_m>}$ and $C_{<n,i_n,j_n>}$ have already been computed, for $0 \leq i_m, j_m, i_n, j_n \leq |S|$.

We see that $C_{<l,i,j>} = \min(C_{<m,i_m,j_m>} + C_{<n,i_n,j_n>} + (i + j) * C(l))$, where the minimum is taken over all i_m, j_m, i_n, j_n that provide valid multicast flows through the node v. Which values of the flows are valid is checked using a reasonably simple algorithm. Due to lack of space, this algorithm has been deferred to [1]. Along with the value $C_{<l,i,j>}$, we store the values of the i_m, j_m, i_n and j_n that resulted in the minimum $C_{<l,i,j>}$. Call these values the *optimal indices* for $C_{<l,i,j>}$. If (i, j) is invalid for link l, $C_{<l,i,j>}$ is set to infinity. Also, if link l is in the to be covered set of links, $C_{<l,0,0>}$ is set to ∞. By proceeding in this fashion from the leaves to the root, we see that we can fill in the entire table.

To complete the algorithm, we attach a virtual link x to the root of N with $C(x) = 0$, and use the same technique to compute $C_{<x,0,0>}$. The minimum cost for covering the given set of required links in N is $C_{<x,0,0>}$. In order to find the actual multicast trees, we first follow the stored optimal indices from $C_{<x,0,0>}$ to the leaves of the tree to determine the actual optimal number of flows in either direction on each link of the tree. Given this information, a simple greedy algorithm finds a set of multicast trees that results in this number of flows. The description of this greedy algorithm is left to the full version of the paper. We present the details of our algorithm, which we call **Tree-Optimal**, in Figure 4.

Theorem 4. *The algorithm* **Tree-Optimal** *finds the optimal cost of a solution to the W-MMTCP in any binary tree in* $O(|E||S|^6)$ *steps.*

The proof is provided in [1]. Following a similar idea, we can also construct an algorithm for solving the S-MMTCP. However, the strong cover requirements make that algorithm somewhat more complicated than the algorithm presented here.

In order to deal with the case that N is a tree of arbitrary degree (instead of just a binary tree), we transform N into a binary tree N'. To do so, choose any node to be the root of the tree N. Then, given a node r with n children, create a virtual node r', make it the child of node r and assign zero cost to the virtual link $\langle r, r' \rangle$. Then pick any $n-1$ other links attached to node r, and move them to node r'. We repeat this process until all nodes are binary; the resulting tree is N'. Since the cost of traveling any virtual link is zero, and no virtual link is in the set of to be covered links, the cost of an optimal solution for the topology N is the same as the optimal cost for the topology N'. Thus, we can find the optimal solution for N by transforming N into the topology N', finding an optimal solution for the topology N', and then transforming the solution for N' into an equal cost solution for N.

We also can deal with the general problem where each multicast tree has a per tree cost. We do this by adding a virtual node u and virtual link $\langle u, v \rangle$ for each source candidate v, and replacing the source candidate v with the virtual node u. We then assign the cost for initializing the multicast tree that was rooted at node v as the cost of the link $\langle u, v \rangle$. The problem then becomes that of finding the optimal cost covering without per tree costs in the resulting network. If we also want to allow the cost of initializing different multicast trees at the same source to vary, we can attach a different virtual node u for each multicast tree rooted at v.

5   Experiments and Findings

In this section, we explore the effectiveness of the heuristics presented in the previous two sections in the context of the Internet2 vBNS backbone network.

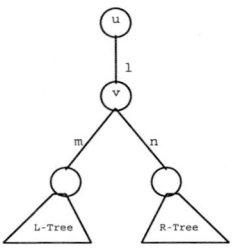

Fig. 3. A Link l Incident to the Same Node as Links m and n.

1. Add a virtual link x whose downstream node is the root of the tree and assign zero cost for traveling link x.
2. Create a table C, with each row labeled with a link l, and each column labeled by $<i,j>$, for $0 \le i,j \le |S|$. Initialize every entry in the table to $+\infty$.
3. For each link l, let SU_l and SD_l be the number of sources above and below l respectively. Let RU_l and RD_l be the number of receivers above and below l respectively.
4. $\forall l_i$, such that l_i is a link attached to a leaf node:
    if $(SD_{l_i} == 1)$ then $C_{<l_i,0,1>} = C(l_i)$
      if $(RD_{l_i} == 1)$ then
          $C_{<l_i,j,k>} = C(L_i) * (j+k), \forall j,k, 0 \le j \le SU_i, 0 \le k \le 1$. endif
    else
        if $(RD_{l_i} == 1)$ then $C_{<l_i,j,0>} = C(L_i) * j, \forall j, 0 \le j \le SU_i$ endif
    if $(l_i \in \mathcal{L})$ then $C_{<l_i,0,0>} = +\infty$ else $C_{<l_i,0,0>} == 0$ endif
5. Choose any link l_i, such that row l_i has not been computed, and l_i is incident to a vertex that is also incident to links l_j and l_k such that rows l_j and l_k have been computed. Let v be the node they share.
    $C_{<l_i,p_3,q_3>} = \min(C_{<l_j,p_2,q_2>} + C_{<l_k,p_1,q_1>} + C(l_i) * (p_3 + q_3))$,
        where $valid(p_3, q_3, p_2, q_2, p_1, q_1, v)$ is true.
    If $(l_i \in \mathcal{L})$ then $C_{<l_i,0,0>} = +\infty$ endif
6. If all the links are done, then stop and return $C_{<x,0,0>}$. Else go to step 5.

Fig. 4. The Algorithm **Tree-Optimal**. This algorithm finds the cost of the optimal solution to the W-MMTCP on a binary tree topology. The function *valid* determines whether or not the integers p_3, q_3, p_2, q_2, p_1, and q_1 define valid sets of flows at the vertex v. Due to lack of space, this algorithm has been deferred to [1].

As this network is sparse, we also consider a set of denser, randomly generated networks.

5.1   The vBNS Network

We consider the vBNS Internet2 backbone network [18]. It maintains native IP multicast services using the PIM sparse-mode routing algorithm. The vBNS multicast logical topology (as of October 25, 1999) is illustrated in [18]. It consists of 160 nodes and 165 edges. The vBNS infrastructure was built out of ATM. Each node represent a physical location and each link represents a physical interconnection between some two routers from different locations. The link bandwidths vary between 45M (DS3) and 2.45G (OC48). Since the more detailed topology within each physical location is not available to us, we treat each node as a router and focus on the logical topology in our experiments. In addition, we assume that the cost of using a link for measurement within one multicast session is inversely proportional to its bandwidth. Last, we assume that only the leaves in the topology (i.e., node of degree one) can be a sender or a receiver.

Tree Heuristic. In section 4 we proposed the algorithm **Tree-Optimal** that is guaranteed to find optimal solutions in polynomial time for any tree topology. We propose and study a heuristic based on that algorithm. This heuristic uses the observation that the topology of a network such as vBNS is very close to a tree. Furthermore, the bandwidth of the small number of links that create cycles tends to be high, and thus presumably have low cost.

The heuristic can be applied to any topology, and proceeds in four phases. In the first phase, the topology is converted to a tree by condensing every cycle into a single super-node. In the second phase, algorithm **Tree-Optimal** is run on the resulting tree. This yields a set of multicast trees, defined by a list of senders, and for each sender a list of receivers. In the third phase, this set of multicast trees is mapped back to the original topology, so that the same set of senders each send to their respective receivers. This is guaranteed to cover all of the required links that were not condensed into super-nodes, but may leave required links that appear in cycles uncovered. The fourth and final phase uses the fast greedy heuristic to cover any such edge.

Note that the cost of the solution obtained by **Tree-Optimal** is a lower bound on the cost of the solution to the actual topology. This implies several important properties of this heuristic. Call any link that appears on a cycle in the graph a *cycle link*. If all of the cycle links have zero cost, and no cycle link must be covered, then the tree heuristic is guaranteed to produce an optimal solution. Also, if there are not many cycle links, or they all have relatively small cost, then the solution found by the heuristic will usually be close to the lower bound, and thus close to optimal. The fact that the heuristic produces this lower bound is also useful, as it allows one to estimate how close to optimal the solution produced by the heuristic is.

Effectiveness of Heuristics. In Section 3 we introduced the greedy algorithm and the 0-1 integer programming algorithm for approximating S-MMTCP and W-MMTCP, and described their worst case approximation ratio bounds. In order to approximate the general MMTCP in polynomial time, we also proposed a fast greedy heuristic in Section 3. In this section we study the average performance of these algorithms and heuristics through experiments on the Internet2 backbone networks. Since the topologies of these networks are close to tree topologies, we include the performance of the tree heuristic on Internet2 backbones in our study.

To create a suite of problem instances, we varied the sizes of the source and receiver candidate sets. In addition, for a particular pair of source candidate set and receiver candidate set, we chose the size of the set of links that must be covered to be proportional to the size of the set of links that the source candidate set and the receiver candidate set can identify. For each problem size, we generated 100 random problem instances for the vBNS multicast network. For each of these problem instances, we determined the cost of the solution found by each algorithm. We assumed that all the multicast trees have the same fixed initialization cost.

We ran the algorithms on inputs where the number of source candidates is eight and the number of receiver candidates varies from eight to sixteen. For small problem instances such as these, the optimal solutions can be computed for these problem sizes using exhaustive search, and this can be used to check the quality of the approximation results. For both W-MMTCP and S-MMTCP, we used the 0-1 integer programming, greedy and fast greedy algorithms to approximate the 100 problem instances on vBNS for each of the problem size. In addition, we used the tree heuristic to approximate the W-MMTCP. We compare the performance of the algorithms for S-MMTCP in Figure 5 and W-MMTCP in Figure 6. In both figures, the ratio of the solutions found by the approximation algorithm to the optimal solutions is plotted. For each approximation algorithm, we sort the ratios in ascending order. Thus, for example, problem instance 1 for each algorithm represents the instance where that algorithm performed the closest to optimal, and may correspond to different inputs for the different algorithms. We present plots for two different problem sizes: 8 sources and 8 receivers, as well as 8 sources and 16 receivers.

In Figure 5, it is surprising to see that the fast greedy algorithm produces the same solution as the greedy algorithm and that the 0-1 integer programming algorithm yields the optimal solution on most inputs when the problem size is small. As the problem size increases, the 0-1 integer programming is less likely to produce the optimal solution and the difference between the fast greedy and the optimal seems to increase slowly.

In the case of approximating W-MMTCP, the tree heuristic out-performs both the greedy algorithm and the fast greedy algorithm on most problem instances. The quality of the 0-1 integer programming algorithm decreases as the problem size increases. The results from the fast greedy are only slightly worse than those from the greedy algorithm. The difference between the fast greedy algorithm and greedy algorithm seems to change very slowly as the problem size increases.

We have also conducted a similar study with similar results on the Internet2 Abilene backbone network [19]. Details of this study are found in [1]

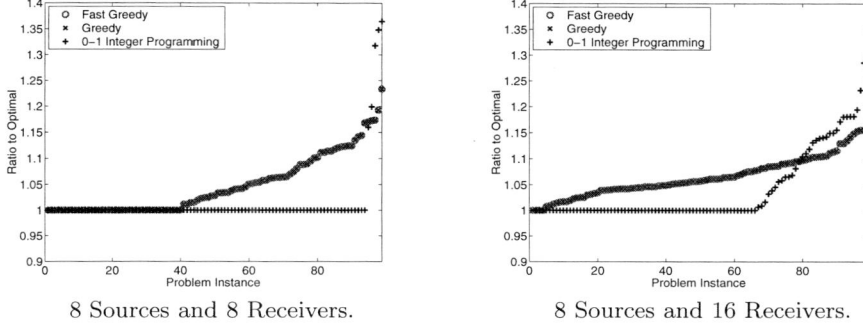

8 Sources and 8 Receivers.   8 Sources and 16 Receivers.

Fig. 5. Comparison of Approximation Algorithms for the S-MMTCP on vBNS.

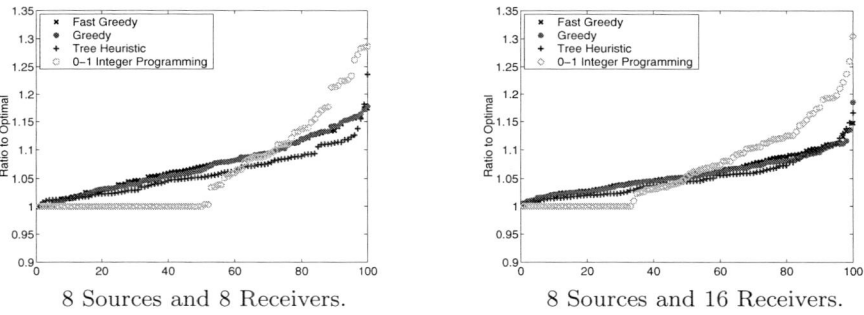

8 Sources and 8 Receivers.   8 Sources and 16 Receivers.

Fig. 6. Comparison of Approximation Algorithms for the W-MMTCP on vBNS.

5.2   Experiments on Dense Networks

vBNS is quite sparse, i.e., it only contains a very small number of additional edges beyond that of a tree topology containing the same number of nodes. In this section, we investigate how our algorithms perform on denser network than vBNS. Unfortunately, we had no such multicast topologies available to us. Instead, we make use of randomly generated topologies.

We generated ten 100-node transit-stub undirected graphs using GT-ITM (Georgia Tech internetwork topology model). For more details about the transit-stub network model, please refer to [17]. The average out-degree is in the range of $[2.5, 5]$. We assigned two costs to each edge in the graphs, one for each direction. These cost are uniformly distributed in $[2, 20]$. By randomly picking the source candidate set, receiver candidate set and to-be covered set of links and then assigning costs to the edges, we generated ten problem instances for each graph. We ran all algorithms on a total of 100 problem instances and compared their performance. We assumed that all the multicast trees have the same fixed initialization cost.

We ran the algorithms on inputs where the number of source candidates is eight and the number of receiver candidates varies from eight to sixteen. We used the 0-1 integer programming, greedy and fast greedy algorithms to approximate

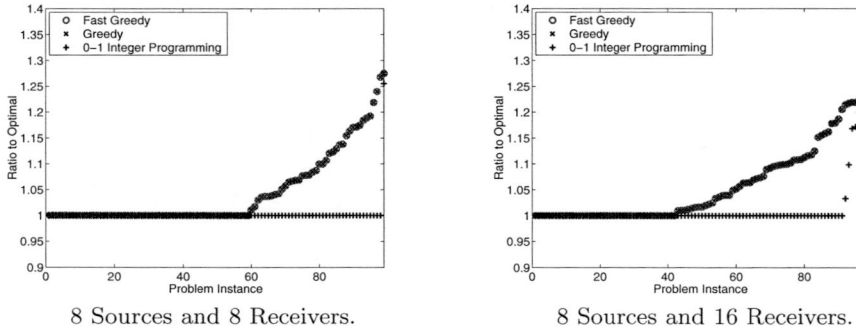

8 Sources and 8 Receivers.   8 Sources and 16 Receivers.

Fig. 7. Comparison of Approximation Algorithms for the S-MMTCP on 100-node transit-stub.

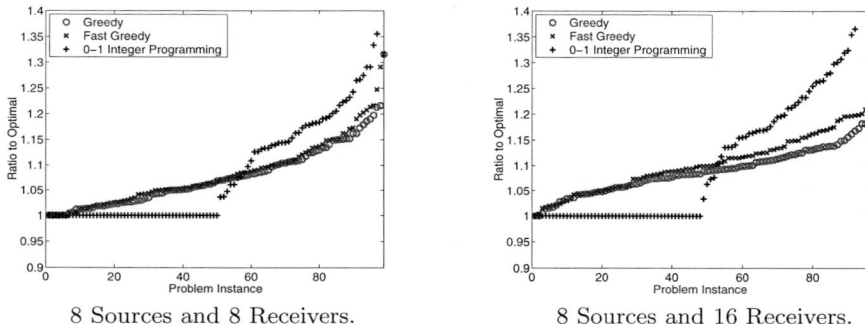

8 Sources and 8 Receivers.   8 Sources and 16 Receivers.

Fig. 8. Comparison of Approximation Algorithms for the W-MMTCP on vBNS on 100-node transit-stub.

the 100 problem instances for each of the problem sizes. We compare the performance of the algorithms for the S-MMTCP in Figure 7 and the W-MMTCP in Figure 8. In both figures, the ratio of the solution found by the approximation algorithms to the optimal solution is plotted. For each approximation algorithm, we sorted the ratios ascendantly.

In Figure 7, it is surprising to see that the fast greedy algorithm produces the same solution as the greedy algorithm and the 0-1 integer programming yields the optimal solution on most inputs. As the problem size increases, the greedy and fast greedy algorithm are less likely to produce the optimal solution.

In the case of approximating W-MMTCP, the 0-1 integer programming algorithm yields the optimal solution on about half the problem instances. However, it yields worse results than the greedy and fast greedy algorithm for about forty percent of the problem instances. The results from the fast greedy are slightly worse than those from the greedy algorithm in most of cases. The difference between the fast greedy algorithm and greedy algorithm seems to increase very slowly as the problem size increases.

6   Conclusions

In this paper we focussed on the problem of selecting trees from a candidate set in order to cover a set of links of interest. We identified three variation of this problem according to the definition of cover and addressed two questions for each of them:

- Is it possible to cover the links of interest using trees from the candidate set?
- If the answer to the first question is yes, what is the minimum set of trees that can cover the links?

We proposed computationally efficient algorithms for the first of these questions. We also established, with some exceptions, that determining the minimum cost set of trees is a hard problem. Moreover, it is a hard problem even to develop approximate solutions. One exception is when the underlying topology is a tree in which case we present efficient dynamic programming algorithms for two of the covers. We also proposed several heuristics and showed through simulation that a greedy heuristic that combines trees with three or fewer receivers performs reasonably well.

References

1. M. Adler, T. Bu, R. Sitaraman, D. Towsley. "Tree Layout for Internal Network Characterizations in Multicast Networks," *UMass CMPSCI Technical Report 00-44.*
2. A. Adams, T. Bu, T. Friedman, J. Horowitz, D. Towsley, R. Caceres, N. Duffield, F. Lo Presti, S.B. Moon, V. Paxson. "The use of end-to-end multicast measurements for characterizing internal network behavior," *IEEE Communications Magazine*, **38**(5)152-159, May 2000.
3. K. Almeroth, L. Wei, D. Farinacci. "Multicast Reachability Monitor (MRM)" IETF Internet Draft, Oct. 1999.
4. K. Almeroth. "The evolution of multicast," *IEEE Network*, **14**(1)10-21, Jan. 2000.
5. R. Chalmers and K. Almeroth, "Modeling the Branching Characteristics and Efficiency Gains of Global Multicast Trees", *IEEE Infocom 2001* Anchorage, Alaska, USA, April 2001.
6. V. Chvátal "A greedy heuristic for the set-covering problem" *Mathematics of Operations Research,* 4(3)233-235, Aug. 1979.
7. M. Coates, R. Nowak. "Network loss inference using unicast end-to-end measurement", *Proc. ITC Conf. IP Traffic, Modeling and Management*, Sept. 2000.
8. S. Deering, D. Estrin, D. Farinacci, V. Jacobson, C-G Liu, L. Wei, "The PIM Architecture for Wide-Area Multicast Routing," *IEEE/ACM Transactions on Networking*, 4, 2, 153–162, April 1996.
9. C. Diot, B.N. Levine, B. Lyles, H. Kassem, D. Balensiefen. "Deployment issues for the IP multicast service and architecture," *IEEE Network*, **14**(1)78-89, Jan. 2000.
10. N.G. Duffield, F. Lo Presti, V. Paxson, D. Towsley. "Inferring Link Loss Using Striped Unicast Probes", to appear in *Proc. INFOCOM 2001.*
11. V. Paxson, "End-to-End Routing Behavior in the Internet". *IEEE/ACM Transactions on Networking*, Vol.5, No.5, pp. 601-615, October 1997.
12. T. Pusateri, "Distance Vector Multicast Routing Protocol", Internet Draft, draft-ietf-idmr-dvmrp-v3-04.ps, Feb. 1997.

13. A. Reddy, R. Govindan, D. Estrin. "Fault isolation in multicast trees," *Proc. SIG-COMM 2000*, Sept. 2000.
14. Y. Shavitt, X. Sun, A. Wool, B. Yener. "Computing the Unmeasured: An Algebraic Approach to Internet Mapping" to appear in *Proc. INFOCOM 2001*.
15. A. Srinivasan "Improved approximation of packing and covering problem " *Proceedings of the Twenty-Seventh Annual ACM Symposium on Theory of Computing* 268-276, Las Vegas, Nevada, June 1995.
16. J. Walz, B.N. Levine. "A practical multicast monitoring scheme," U. Massachusetts Computer Science Technical Report 30-2000, June 2000.
17. Ellen W. Zegura, Ken Calvert and S. Bhattacharjee " How to model an internetwork" *Proceedings of IEEE Infocom '96, San Francisco, CA.*
18. http://www.vbns.net/netmaps/multicast.html.
19. http://www.abilene.iu.edu/images/ab-mcast.pdf.